FOREST WILDLIFE ECOLOGY AND HABITAT MANAGEMENT

FOREST WILDLIFE ECOLOGY AND HABITAT MANAGEMENT

DAVID R. PATTON

CRC Press
Taylor & Francis Group
Boca Raton London New York

CRC Press is an imprint of the
Taylor & Francis Group, an **informa** business

CRC Press
Taylor & Francis Group
6000 Broken Sound Parkway NW, Suite 300
Boca Raton, FL 33487-2742

First issued in paperback 2019

ISBN-13: 978-1-4398-3702-3 (hbk)
ISBN-13: 978-0-367-38354-1 (pbk)

Library of Congress Cataloging-in-Publication Data

Patton, David R.
  Forest wildlife ecology and habitat management / David R. Patton.
    p. cm.
  Includes bibliographical references and index.
  ISBN 978-1-4398-3702-3
  1. Animal ecology. 2. Forest animals. 3. Forest ecology. 4. Wildlife habitat improvement. I. Title.

SK356.W54P38 2010
639.90915'2--dc22   2010015644

Contents

Preface... xiii
Acknowledgments..xvii
About the Author xix

Chapter 1   Working Concepts 1

    1.1   Introduction 1
    1.2   Common Knowledge Axioms 1
    1.3   Wildlife in a Forest..3
           1.3.1   What Is Wildlife? 3
           1.3.2   Classification 4
           1.3.3   Life Form..5
           1.3.4   Life History of Animals 5
    1.4   Environment of Forest Wildlife..5
           1.4.1   The Centrum and Web 6
           1.4.2   What Is Habitat?...7
    1.5   Plants in a Forest 10
           1.5.1   Life Form.. 10
           1.5.2   Life History of Plants 11
    1.6   Water in a Forest.. 11
           1.6.1   Water Properties 12
           1.6.2   Freshwater Habitats 12
           1.6.3   Lakes 13
           1.6.4   Streams.. 14
           1.6.5   Seeps and Springs 14
           1.6.6   Riparian Zones 15
    1.7   Forest Diversity 16
           1.7.1   Horizontal and Vertical Diversity 17
           1.7.2   Layers 17
           1.7.3   Components.. 18
           1.7.4   Arrangement of Components 20
           1.7.5   Biological Diversity...20
           1.7.6   Edge...23
    1.8   Vegetation 24
           1.8.1   Classification 25
           1.8.2   Stands of Trees 28
           1.8.3   Openings in Stands 28
           1.8.4   Dead Trees and Snags 28
           1.8.5   Coarse Woody Debris 29

1.9   Forests As Ecological Systems...30
        1.9.1   Hierarchical Organization.....................................30
        1.9.2   Structure and Processes 31
        1.9.3   Nutrient Cycles..32
        1.9.4   Energy Flow 33
        1.9.5   Nutrient and Energy Pathways 34
        1.9.6   Soil...35
        1.9.7   Time and Space 36
        1.9.8   Ecological Amplitude..36
        1.9.9   Limiting Factors...37
1.10   Forest Succession...37
        1.10.1   Agents of Change 38
        1.10.2   Primary Succession...38
        1.10.3   Secondary Succession 39
        1.10.4   Tree Structural Stages 39
        1.10.5   Old Forest Conditions...41
        1.10.6   Changes over Time..42
1.11   Healthy Forest Conditions 44
        1.11.1   Long-Term Productivity 44
        1.11.2   Natural Events 44
        1.11.3   Sustainability...45
        1.11.4   Climate Change..45
1.12   The Menominee Paradigm 47
1.13   The Principle of Complementarity..48
Knowledge Enhancement Reading 49
References 49

Chapter 2   Survive and Reproduce 55
2.1   Introduction 55
2.2   Hazards...55
        2.2.1   Fire 55
        2.2.2   Roads..57
        2.2.3   Fences...58
        2.2.4   Weather...58
2.3   Diseases 59
        2.3.1   Disease-Producing Agents 59
        2.3.2   Pesticides, Herbicides, and Insecticides 61
2.4   Predators...61
        2.4.1   The Leopold Report 61
        2.4.2   Lotka-Volterra Equations 62
        2.4.3   General Propositions 63
2.5   Humans...64
        2.5.1   Habitat Destruction 64
        2.5.2   Illegal Hunting 65

2.6   Biology...66
       2.6.1   Reproduction 66
       2.6.2   Behavior 68
       2.6.3   Territory..68
       2.6.4   Ecological Niche 69
       2.6.5   Population Characteristics...69
       2.6.6   Uninhibited Growth Model 73
       2.6.7   Rates of Population Change 74
       2.6.8   Inhibited Growth Model..78
       2.6.9   Carrying Capacity 81
       2.6.10 Life Table...82
       2.6.11 Fluctuations 83
       2.6.12 Demographic Vigor 85
2.7   Resources..85
       2.7.1   Food..86
       2.7.2   Cover 88
       2.7.3   Water 89
       2.7.4   Space 91
       2.7.5   Mobility..91
       2.7.6   Order of Magnitude 93
Knowledge Enhancement Reading 95
References 95

Chapter 3   Integrating Forestry and Wildlife Management............................ 101

3.1   Introduction 101
3.2   Forest Types... 101
       3.2.1   Eastern Forest Types 103
               3.2.1.1   White-red-jack pine (Type 10)................. 103
               3.2.1.2   Spruce-fir (Type 11)................................. 106
               3.2.1.3   Longleaf-slash pine (Type 12) 106
               3.2.1.4   Loblolly-shortleaf pine (Type 13)............ 108
               3.2.1.5   Oak-pine (Type 14) 108
               3.2.1.6   Oak-hickory (Type 15)............................. 109
               3.2.1.7   Oak-gum-cypress (Type 16) 109
               3.2.1.8   Elm-ash-cottonwood (Type 17) 109
               3.2.1.9   Maple-beech-birch (Type 18).................... 110
               3.2.1.10   Aspen-birch (Type 19) 110
       3.2.2   Western Forest Types 110
               3.2.2.1   Douglas-fir (Type 20)............................... 110
               3.2.2.2   Hemlock-Sitka spruce (Type 21)............... 111
               3.2.2.3   Ponderosa pine (Type 22) 111
               3.2.2.4   Western white pine (Type 23)................... 112
               3.2.2.5   Lodgepole pine (Type 24)......................... 112
               3.2.2.6   Larch (Type 25) 112

	3.2.2.7	Fir-spruce (Type 26)	112
	3.2.2.8	Redwood (Type 27)	113
	3.2.2.9	Pinyon-juniper (Type 28)	113
	3.2.2.10	Western hardwoods (Type 29)	114

3.3 Forest Management 114
    3.3.1 Silvicultural Systems.. 115
    3.3.2 Tree Species Tolerance.. 115
    3.3.3 Even- and Uneven-Aged Stands 116
    3.3.4 Clearcutting 116
    3.3.5 Shelterwood.. 119
    3.3.6 Seed Tree.. 120
    3.3.7 Selection 120
    3.3.8 Old Forest Conditions.. 120
    3.3.9 Thinning.. 121
    3.3.10 Spacing of Trees 121
    3.3.11 Forest Regulation 121
    3.3.12 Rotation 122
    3.3.13 Cutting Cycle... 123
3.4 Wildlife Management.. 124
    3.4.1 Hazards... 124
    3.4.2 Diseases.. 125
    3.4.3 Humans 127
    3.4.4 Predators... 127
    3.4.5 Biology 128
    3.4.6 Resources 134
3.5 Integrating Factors.. 136
    3.5.1 A Matter of Scale 137
    3.5.2 Time 137
    3.5.3 Natural Rotation 137
    3.5.4 Stand Size and Distribution.. 138
    3.5.5 Log-Normal Distribution... 138
    3.5.6 Nonadjacency Constraint 139
    3.5.7 The Landscape Factor 140
    3.5.8 Tradition and Other Considerations 141
    3.5.9 Grazing Systems... 143
    3.5.10 Conflicts and Competition.. 143
    3.5.11 National Research Council.. 145
3.6 Restore and Maintain Resources and Diversity 146
    3.6.1 Wildlife Coordination and Restoration Measures..... 151
    3.6.2 Timber Coordination and Restoration Measures 153
    3.6.3 Range Coordination and Restoration Measures....... 155
    3.6.4 Engineering Coordination and Restoration Measures.. 156

3.6.5   Watershed Coordination and Restoration Measures 157

3.6.6   Recreation Coordination and Restoration Measures 157

3.6.7   Land Use Coordination and Restoration Measures 158

3.6.8   Aquatic Coordination and Restoration Measures 158

Knowledge Enhancement Reading 159

References 160

Chapter 4   Management Strategies 163

4.1   Introduction 163
4.2   Ecosystems 163
     4.2.1   Local and Formational 164
     4.2.2   Landscapes 164
     4.2.3   Habitat Area 164
4.3   Wildlife Habitat Relationships 166
     4.3.1   Single Species 168
     4.3.2   Multispecies 169
     4.3.3   Featured Species 170
     4.3.4   Guilds 171
     4.3.5   Life Form 171
4.4   Ecological Factors 172
     4.4.1   Indicators 173
     4.4.2   Biological Diversity 174
     4.4.3   Edge 174
     4.4.4   Habitat Corridors 176
     4.4.5   Gap Analysis 177
4.5   Habitat Models 177
     4.5.1   Factors to Consider 178
     4.5.2   Decision Making 180
         4.5.2.1   Decision Tree 181
         4.5.2.2   Expert Systems 181
     4.5.3   Pattern Recognition 183
     4.5.4   Habitat Indices 185
     4.5.5   Stand Structure Model 186
     4.5.6   Wildlife Data Models 186
4.6   Habitat Evaluation and Monitoring 188
     4.6.1   Ecosystem Functioning 188
     4.6.2   Habitat Quality and Populations 189
     4.6.3   Habitat Suitability Index 189
     4.6.4   Condition and Trend 190
     4.6.5   Monitoring Habitat Change 194

4.7   Integrated Management.. 194
       4.7.1   Adaptive Management... 195
4.8   Planning.. 196
       4.8.1   Laws and Policy 197
       4.8.2   Fish and Wildlife Coordination Act........................ 197
       4.8.3   Multiple-Use, Sustained-Yield Act......................... 198
       4.8.4   Sikes Act... 198
       4.8.5   National Environmental Policy Act 198
       4.8.6   Endangered Species Act.. 198
       4.8.7   Forest and Rangeland Renewable Resources
                Planning Act (RPA)... 199
       4.8.8   National Forest Management Act (NFMA) 199
       4.8.9   Governing Policies 199
       4.8.10   Why Plan? 201
       4.8.11   Factors to Consider... 201
       4.8.12   Operational Plans 202
       4.8.13   Checklist for Background Information 203
       4.8.14   Decision Support Systems....................................... 203
       4.8.15   Project Implementation 204
4.9   Inventory.. 205
       4.9.1   Multiresource Inventory.. 205
       4.9.2   Categories of Inventory 206
       4.9.3   Intensity of Inventory 207
       4.9.4   Map Scale.. 208
       4.9.5   Geographic Information Systems............................ 209
4.10   Sampling... 209
       4.10.1   Types of Sampling.. 210
       4.10.2   Shape and Size of Sample Units............................. 211
       4.10.3   Sampling Design 212
       4.10.4   Information to be Collected 213
       4.10.5   Uses of Habitat Data... 214
       4.10.6   Estimating Animal Numbers 215
       4.10.7   Counting Animals (Direct) 215
       4.10.8   Counting Signs (Indirect)....................................... 216
       4.10.9   Tree Measurements 218
Knowledge Enhancement Reading 220
References 220

Chapter 5   Forest Attributes and Wildlife Needs (FAAWN) 225

5.1   Background Information 225
5.2   Relational Database Systems.. 228
       5.2.1   Structure and Organization 228
       5.2.2   Quality Control... 228
       5.2.3   Relation Theory.. 228
       5.2.4   Relationship Systems.. 229

		5.2.5	Design Criteria	230
		5.2.6	How Relational Systems Work	231
		5.2.7	Rules for Placing the Links	231
		5.2.8	Determining the Links	232
		5.2.9	Evaluate and Test the Design	232
	5.3		Developing a Relational Data Model	233
		5.3.1	Animals	243
		5.3.2	Food	245
		5.3.3	Life-Needs	245
		5.3.4	States	246
		5.3.5	Hypothesized	247
		5.3.6	Regions	247
		5.3.7	Types-States	248
		5.3.8	Types-Codes	248
		5.3.9	Plants-Types	248
		5.3.10	Plants	248
		5.3.11	National	248
	5.4		A Way of Learning	248
		5.4.1	Using the Data Model	248
		5.4.2	Examples of Output	249
		5.4.3	Suggested Additions	249
		5.4.4	Entering and Editing Data	250
	5.5		Using the Computer Disk	250
			Knowledge Enrichment Reading	251
			References for Validating Information	251
			References	253

Appendix: Conversion Factors 255

Index 257

Supplementary Resources Disclaimer

Additional resources were previously made available for this title on CD. However, as CD has become a less accessible format, all resources have been moved to a more convenient online download option.

You can find these resources available here: www.routledge.com/9781439837023

Please note: Where this title mentions the associated disc, please use the downloadable resources instead.

Preface

When I was a young boy in West Virginia, my backyard was a forest. This forest was my classroom; learning the "way of the woods" from traditional knowledge of local people, passed on by generations of settlers in the Appalachian Mountains, was a way of life. During these formative years, I began to store mental images of how plants and animals were associated with certain forest characteristics. These images became patterns that were replications of combinations that were observed in other areas. It was these life experiences that had a strong influence on my choice of profession and how I would later use my early experiences and accumulated knowledge for practical application.

My undergraduate and graduate education focused on forestry, wildlife, fisheries, watersheds, and systems analysis. The first four topics came naturally, but systems analysis did not enter my thoughts until I read a publication by George Van Dyne (1966), in which he proposed that a systems approach had a lot to offer in ecology, especially in renewable resource management, and that influenced the direction of my future graduate studies and research.

In reading a book by Peter Klopfer (1969), I was introduced to the term *Umwelt*, and this created an interest in wanting to learn more about how animals viewed, sensed, or selected food and cover components as part of a forest system. A third publication that influenced my thinking was *The Ecological Web* by Andrewartha and Birch (1984), in which they discuss the theory of environment, which is really all about ecological systems (ecosystems). The three publications mentioned have been a major source of ideas for my research and teaching.

In my first forestry job on the Cleveland National Forest, I began to understand how important it was to integrate forestry, wildlife, range, and recreation in a broader context even though I had to make decisions on day-to-day operations at local levels. Later, a position on the Santa Fe National Forest provided experiences in interacting with the forest timber staff to have wildlife considered in their harvesting plans by adjusting sizes and shapes of cuts and leaving protective cover, even though there was no policy or directive to guide my actions. While I was not always successful, the process was established for possibilities in future confrontations. It also was an opportunity to integrate forestry and wildlife objectives into harvesting plans under the principle of multiple use. As a forest biologist, I began to see in a direct way that what was done at a local area had an effect on adjacent areas in larger management units, which at that time were called compartments and working circles but in reality were human-defined ecosystem units and landscapes.

The position of project leader in the U.S. Forest Service provided experience in several landscape-level studies in the Beaver Creek Watershed on the Coconino National Forest, the Apache National Forest, and the Fort Bayard research site on the Gila National Forest. All these studies were designed to determine the effects of timber harvesting and range management practices on different species of wildlife. During my research assignment, the opportunity occurred to work with the U.N. Food and Agriculture Organization to conduct an analysis of a large river basin in the Luangwa Valley, Zambia. This change in environment provided the experience of working at a very large landscape level to delineate areas for different purposes, such as game management units, tourism, and national parks. It was also during this assignment that it became clear that human wants and needs are the same worldwide, but the circumstances where one lives are different.

Considerable time on the Luangwa project was devoted to aerial surveys to document elephant numbers and locations. In the Luangwa Valley, the elephant is an ecological dominant that had to be managed in large landscapes that included forests, grassland plains, and riverine vegetation. As a wildlife biologist and forester, it was a rewarding experience to immobilize an elephant and put a "plate-sized" tag in his ear with the number 33 (my age at the time). Marking elephants was necessary to determine their daily and seasonal movements to delineate areas for potential national parks and areas for controlled hunting. I named the elephant Chifungwe, and he later became the subject of a novel of my life as a wildlife biologist in Africa during a time of political and social change following independence.

Working and living in foreign countries confirmed my belief that forests, with their biological attributes, follow the same laws of nature and have similar characteristics whether in a deciduous forest in the United States, mopane and brachystegia woodlands in Zambia, riverine acacia in Kenya, the wet banana belt on Kilimanjaro in Tanzania, or the tableland rain forests of Australia. The overriding influence in forests is the volume (size and density), distribution, and combinations of trees as habitat and the by-products of trees that become the debris on the forest floor and habitat for small animals.

My approach to forest wildlife ecology at the management, research, and academic levels has been to formulate conceptual models as a way to categorize and understand complex relationships in systems and to identify areas of overlap that could be habitat components for several species. This approach to analyzing data relates to my education and experience as a practicing forester and biologist, which required me to use and later produce information for field application.

In the biological and ecological science fields, there are definitions that are sometimes generalizations to explain complex situations or activities. Concept definitions, such as diversity, juxtaposition, and interspersion, are important intellectual considerations, but sometimes they may not have a direct management application because they lack a quantitative or subjective measure (can be seen but not easily described). However, they are useful tools that can enhance and broaden the understanding of complex management situations. Some of the concept definitions in this textbook are works or hypotheses in progress and will change over time as experience in their use accumulates. For ease in reading, and unless otherwise noted, all measurements are in the English system of inches, acres, and so on. English-to-metric conversion formulas for common measurements are in the Appendix. Throughout the five chapters, abbreviations are used for some common forest and wildlife factors. These abbreviations are codes used to integrate forest and wildlife habitat structure and components into a relational data model in Chapter 5.

This textbook is not all inclusive of topics in the forestry and wildlife professions. It is intended for undergraduate forestry students and practicing foresters who need a background in wildlife and for undergraduate wildlife students and practicing biologists who need a background in forestry. The topics covered do not replace the forestry and wildlife textbooks, which contain more detail on selected topics. The Knowledge Enhancement Readings sections direct readers to publications that provide expanded detail and in some cases a different explanation or opinion.

The first four chapters of *Forest Wildlife Ecology* are modules that can be used in a semester undergraduate course and are organized for topic assignments and testing. The fifth chapter is intended to encourage readers and students interested in forest habitat relationships by providing a detailed example of how information can be organized and used in a data model. While Chapter 5 does not have to be required as part of a semester course, students and working professionals can use it as a learning tool for individual study. The design of the forest attributes and wildlife needs (FAAWN) model has evolved over many years and is presented for consideration for use by individuals for their own work projects or educational benefit; it is an expanded system previously used by wildlife biologists in the Forest Service. There are sufficient data in each of the relations (tables) to obtain a good understanding of how a relational database management system works. Users of FAAWN can add their own data to the current database or copy the database structure and enter data for a local area.

As a thought for forest and wildlife sustainability for the future, my favorite all-time quotation is from the famous cartoon philosopher Pogo, who said: "I met the enemy and he was us" (Walt Kelly).

May the forests be with you.

David R. Patton
Professor and dean emeritus
Northern Arizona University

REFERENCES

Andrewartha, H. G., and L. C. Birch. 1984. *The ecological web: More on the distribution and abundance of animals.* University of Chicago Press, Chicago.

Klopfer, P. H. 1969. *Habitats and territories: A study of the use of space by animals.* Basic Books, New York.

Van Dyne, George M. 1966. Ecosystem, systems ecology, and systematic ecologies. U.S. Atomic Energy Commission, Oak Ridge National Laboratory, Oak Ridge, TN.

Acknowledgments

For 50 years, I have enjoyed discussions with people in the forestry and wildlife profession, and my thinking and writing has been influenced by many field ecologists. These ecologists could look at complex issues and rephrase them in terms practicing managers could understand and apply in a field situation. Tom Evans, chief Forest Service biometrician and a member of my graduate committee at Virginia Tech, influenced my approach to statistical analysis and presentation of data by directing my thoughts to the use of descriptive and correlative statistics and frequency distributions with Ockam's razor a major controlling factor. "Uncle" Henry Mosby, who was the editor of the first *Wildlife Techniques Manual*, constantly reminded me that habitat was the primary key in managing wildlife, and he was always a source of encouragement for getting management experience first before going into research and the academic community.

Both in the United States and abroad, I have had the pleasure to personally know and discuss common concerns with, work with, or take classes from the following people who have made contributions to the profession and enriched my forestry and wildlife experience: Ken Parker, Thane Riney, Henry Mosby, Burd McGinnes, Tom Ripley, R. L. Smith, Robert Latimore, Hudson Reynolds, Johnny Uys, Roger Hungerford, Simon Seno, Maurice Brooks, Gerry Cross, Lyle Sowles, D. I. Rasmussen, Lloyd Swift, Dale Jones, Bill Zeedyk, Jack Thomas, Jim Klemmedson, Pete Ffolliott, Joyce Berry, Jack Dieterich, Greg Goodwin, Jack Adams, Ward Brady, Elbert Reid, Erwin Boeker, David Brown, Paul Patton, John Bailey, John Gordon, Wally Covington, Norman Johnson, Jerry Franklin, John Sessions, Bill Banbridge, Bill Burbridge, Donald Dodds, John Hall, Keith Menasco, Neil Payne, Lou Stanley, Lawrence Temple, John Lemkuhl, Tom Skinner, Cecelia Overby, Tom Ratcliff, Bob Vahle, Rick Wadleigh, O. C. Wallmo, and Howard Hudak, to name a few.

I owe a debt of gratitude to many Forest Service and state biologists with whom I worked during the time of the Resource Planning Act Assessment; for many years, they would contact me for information. In many cases, I did not have what they needed but the need to do a literature search to provide answers helped me understand what their on-the-ground problems were and to identify potential areas for research. I hope these biologists will recognize their contributions in some parts of *Forest Wildlife Ecology* and will find satisfaction in knowing their ideas were always a major part of my ongoing education.

Several anonymous reviewers provided comments that helped me clarify topics and improve the content. A special thanks is due to Jim Allen from Northern Arizona University and John Bailey from Oregon State University for their review of draft material, to Lisa Kiersley for her expertise in computer graphics, to Greg Brandenburg for his help in information technology and Microsoft software, and to Doris Patton for making my writing readable.

Finally, I would like to thank Randy Brehm, Andrea Grant, Stephanie Morkert, and Marsha Hecht of Taylor & Francis Publishing for their attention to detail and ability to detect where slight changes would strengthen an idea or thought that would make a manuscript better and to Scott Shamblin for his excellent cover design that really captures the content of this book.

David R. Patton

About the Author

A forest wildlife ecologist, professor, and dean emeritus at Northern Arizona University, David Patton holds a bachelor's degree in forestry from West Virginia University, a master's in wildlife management from Virginian Tech, and a PhD in watershed management from the University of Arizona. He is a certified wildlife biologist and has been a member of the Society of American Foresters and the Wildlife Society for 50 years.

After receiving his undergraduate degree in forestry, he began a career in federal, state, private, and international organizations, involving management experience on two national forests; was a research biologist and project leader for the U.S. Forest Service, professor and dean at Northern Arizona University, wildlife expert in Zambia for the United Nations, director of the Center for Wildlife Management Studies in Kenya, and director of the Center for Tropical Rainforest Studies in Australia. He made many trips to Mexico while a Forest Service research biologist.

David has published over 125 manuscripts in refereed journals, in government publications, and in symposium proceedings; in 2004, he received the Gerald Cross Alumni Leadership Award from the College of Natural Resources at Virginia Tech. Other awards include the Gulf Oil National Conservation Award for exceptional service in the cause of conservation; Teacher of the Year, NAU School of Forestry; Professional of the Year, Arizona State Chapter of the Wildlife Society; and several awards for research leadership and productivity by the U.S. Forest Service. He has been associate editor of *Forest Science*, has worked with wildlife professionals on Indian reservations and national forests, and has made consulting and educational trips to the countries of Malawi, Eritrea, Norway, Scotland, England, Wales, Ghana, Tanzania, China, and Costa Rica.

David is an Air Force veteran, lives in Flagstaff, Arizona, and is on the board of directors of FaithWorks, a nonprofit organization that supports programs to help the elderly, sick, and poor in several countries in East Africa, Mexico, Russia, and individual states and Indian reservations in the United States.

1 Working Concepts

1.1 INTRODUCTION

A *forest* is a biological community dominated by trees. When we enter a forest, we first see a physical structure and composition of tall and short plants taking up space in various configurations of openness or tightness. In this setting, we most likely will observe some very busy birds and small mammals letting us know that we are intruders in their environment. As we continue walking through the area, we might sense that this forest is a place of calmness and serenity where nothing is happening. But, masked by the physical environment is a complex system of energy flow and nutrient cycling finely tuned to maintaining a state of equilibrium to local conditions of competition, death, and renewal—unseen but taking place in our presence. Some trees may be in the process of becoming snags, and several snags may have already become decaying logs. In the decomposing residue, there may be a young seedling getting established that will start the birth, death, and decay process again.

Many internal and external factors affect forest change and sustainability over time; therefore, a forest is not a static system. Change occurs as a result of biological, chemical, and physical processes acting singly or in combination and from natural or human-caused disturbances (Figure 1.1). These processes are the primary factors responsible for the successive replacement of one plant community with another. Given that when other environmental factors are held constant, plant communities will determine which wildlife species will be present in time and space, and it is all these connecting factors that are topics in the five chapters in this book.

1.2 COMMON KNOWLEDGE AXIOMS

Since the mid-1950s, considerable information from research has accumulated in the ecological literature from articles in the *Journal of Ecology, Forest Science, Journal of Wildlife Management*, and *Journal of Forestry* and from government publications and textbooks about environment, habitat, and wildlife. From the accumulated literature and from years of knowledge passed on by generations of practicing managers, there is a body of evidence that can be easily translated into ecological axioms. An *axiom* is a statement of self-evident truth that is so obvious it cannot be made clearer. Clarifying concepts was one of the great traits of the early field ecologists, such as Aldo Leopold, Charles Elton, Arthur Bent, Ralph King, Victor Shelford, and Olas Murie, to name just a few.

For use in forest wildlife management, the following axioms were developed from common knowledge to guide the thinking and actions of managers in making decisions and for their use in the planning and assessment process. These nine axioms

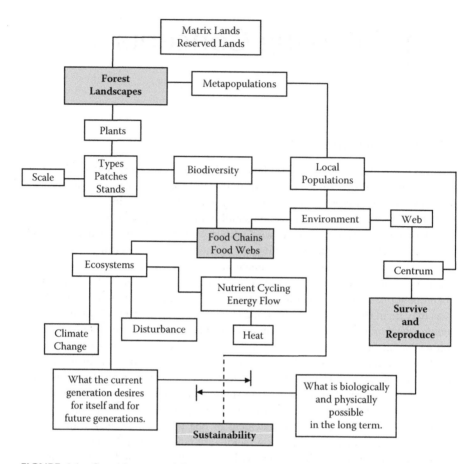

FIGURE 1.1 Cognitive map of factors and issues in a forest that affect the ability of an animal to survive and reproduce.

are presented so that they can be referenced throughout the text in other units when discussing the various topics on environment, habitat, and wildlife. The axioms are not all inclusive, and others could be developed to be more specific in regional and local areas.

Axiom 1. All *plant species*, to survive and reproduce, are directly dependent on energy from the sun and nutrients from soil, air, and water from their surrounding environment.

Axiom 2. All *animal species* are directly or indirectly dependent on plants and plant communities for their energy to survive and reproduce.

Axiom 3. Plant communities change over ecological time under the influence of soil, climate, topography, and organisms and in present time as a result of natural disturbances or human activities.

Axiom 4. Animal species possess different tolerances (ecological amplitudes) to environmental factors; therefore, composition of animal species will change as plant communities change.

Axiom 5. Animals with wide ecological amplitudes (habitat generalists) are less affected by changes in plant communities than animals of low ecological amplitudes (habitat specialists).

Axiom 6. Animal species are integral parts of ecosystem functioning, providing pathways for nutrient cycling and energy flow through food chains and webs.

Axiom 7. Humans are ecological dominants with the ability to make environmental changes that speed up, slow down, alter, or completely eliminate plant succession.

Axiom 8. In the minds of humans, not all plants or animals have equal value; some will be favored at one time and others at another time.

Axiom 9. No ecosystem is completely stable, and change is the rule rather than an exception.

1.3   WILDLIFE IN A FOREST

This book is about wildlife, a subject of many connotations and interpretations, depending on one's particular educational background, experience, and philosophy about living things. For clarity and continuity, it is necessary to define wildlife so that no doubt exists regarding what is meant. *Wild*, in the dictionary, is defined as unrestrained or free roaming. *Life* refers to a quality that distinguishes a functional integrated being from ordinary chemical matter. In this context, *life* suggests either plants or animals, and in some instances *wildlife* refers to both.

Natural resource agencies define wildlife in a variety of ways depending on the specific laws and regulations under which they operate. In many agency definitions, fish are viewed as separate from wildlife, while in a few cases, such as endangered species, invertebrates are often included (Axiom 8). Therefore, a discussion of wildlife in any particular instance requires some terms of reference.

1.3.1   What Is Wildlife?

In North America, the working definition of *wildlife* generally refers to all species of amphibians, birds, fishes, mammals, and reptiles occurring in the wild. *In the wild* is interpreted as being undomesticated and free roaming in a natural environment. For use in forestry, *wildlife* is defined as all undomesticated animal life (Helms 1998).

Forest Wildlife species are those whose primary habitat contains trees as the dominant plant life form. (Patton 1992)

The definition of wildlife given must be applied to the species as a whole. For example, while one deer in a zoo may not be able to roam free, it is still considered wildlife because deer as a species occurs unrestrained in the wild.

Deer in an enclosure would also be considered wildlife even though their movements might be restricted to a specific area. But, what about mice and rats that occur in buildings in a city? These city dwellers are not considered wildlife because they are defined as pests living in a man-made environment. No matter what definition is given, there will always be classes of animals or situations that do not conform.

For this reason, it is desirable to state a definition for the particular circumstances, whether it is for management planning or environmental impact statements.

1.3.2 CLASSIFICATION

The classification of wildlife into categories for management purposes is a continuing process. Many species can be included in several different groups depending on the purpose of the grouping, such as

- Threatened and endangered
- Game, such as small game, big game, waterfowl, and fish
- Furbearer
- Nongame, such as songbirds, predators, and fish
- Nonprotected species
- Introduced, exotic, or feral

A second classification, and one of the most common, is by land use category. This classification usually has the following categories:

- Farm, such as cottontails and quail
- Forest, such as ruffed grouse, black bear, white-tailed deer, and wild turkey
- Wilderness, for example, elk, moose, and grizzly bear
- Wetland, such as species of duck, geese, and muskrat
- Rangeland, such as bison and pronghorn

Another system is that of associating animals with different habitat requirements based on vegetation succession stages (Young and Giese 2003) as follows:

- Early successional species (moose, white-tailed deer)
- Mixed-age forests (snowshoe hare, elk, ruffed grouse)
- Old-growth forests (spotted owl, woodland caribou)
- Edge habitats (raccoons, fox, jays)
- Riparian zones (mink, beaver)

In a scheme to classify all North American mammals and birds according to how closely they are associated with forested areas, Yeager (1961) grouped species into three categories:

- Primarily forest or brushland (714 species)
- Secondarily forest or brushland (195 species)
- Woody cover not or only rarely used (318 species)

The assignment to each category was arbitrary, and in some cases, such as for cottontails, a clearcut designation was difficult because of their wide ecological amplitude.

1.3.3 LIFE FORM

Wildlife life form follows the scheme of taxonomic classification as *amphibians, birds, fish, mammals,* and *reptiles.* In some cases, *insects* are included, especially when developing habitat relationships to include all animal species. Other authors have used the term *life form* in different ways: Barbour et al. (1999) for plant "growth form" and Thomas (1979) for wildlife breeding and feeding activities. There should not be any confusion concerning the use of the term in different scenarios because the context of the discussion will indicate which definition is appropriate. The use of life form for animals (amphibian, bird, fish, invertebrate, mammal, or reptile) is based on physiognomic characteristics, the same as it is for plants.

1.3.4 LIFE HISTORY OF ANIMALS

Life history information for animals describes all the events that happen from birth to the death of an individual. It includes the specific information on reproduction as well as food and cover requirements. Life history for wildlife is often separated into two categories, reproduction and habitat requirements, depending on the intended use. A complete life history includes distribution data as well as taxonomic classification. In documenting life history, the major factors to include are taxonomic information, physical description, distribution, breeding characteristics, behavior, and resources (food, cover, water, space).

Life history information can be presented in great detail, such as in *White-Tailed Deer: Ecology and Management* (Halls 1984), or in a summary form, providing facts necessary to become familiar with the species or use in a data model, such as

Common name:	White-tailed deer
Scientific name:	*Odocoileus virginianus*
Breeding season:	September–February
Mating behavior:	Polygamous
Young per season:	One or two
General habitat:	Coniferous and deciduous forest with shrub understory
Food habits:	Browse, forbs, grasses

Other examples of detailed life history accounts include *Elk of North America* (Thomas and Toweill 1982), *The Desert Bighorn Sheep in Arizona* (Lee 1989), and *Ruffed Grouse* (Atwater and Schnell 1989).

1.4 ENVIRONMENT OF FOREST WILDLIFE

The earth is our global environment; a forest has an environment, and locally, humans have an environment in homes, in workplaces, and surrounding their daily activities. Since the 1960s, some people have come to call themselves environmentalists. So, what is this abstract but pervasive factor that so affects our daily lives and, more specifically, those animals defined in this textbook as forest wildlife?

1.4.1 THE CENTRUM AND WEB

Insight into the environment and its effects on animal populations resulted in a publication, *The Ecological Web* (Andrewartha and Birch 1984), that explained environment in a comprehensive theory:

> The environment of an animal consists of everything that might influence its chance to survive and reproduce. (p. 7)

To support their theory of environment Andrewartha and Birch formulated three propositions:

1. Any environment is made up of a *centrum* of *directly* acting components and a *web* of *indirectly* acting components.
2. The centrum of the environment of any animal has four divisions; each division houses a characteristic set of components: resources, mates, malentities (decimating factors), and predators.
3. The web is comprised of a number of systems of branching chains. A link in the chain may be a living organism or its artifact, inorganic matter, or energy.

The four directly acting components or factors (Proposition 2) can be interpreted more broadly by *six factors* that affect the chance for an animal to survive and reproduce: biology, humans, diseases, predators, hazards, and resources (Patton 1992). These six factors are in the centrum (Figure 1.2) and are stated as Conceptual Equation 1.1:

$$PR_{SR} = f (Bi + Hu + Di + Pr + Ha + Re) \qquad (1.1)$$

where PR_{SR} is the probability that an individual or population will survive and reproduce, and where f is a function of biology (Bi), humans (Hu), disease (Di), predators (Pr), hazards (Ha), and resources (Re).

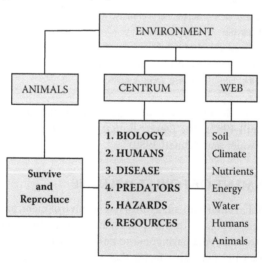

FIGURE 1.2 Factors in the environment that affect the ability of an animal to survive and reproduce.

Five of the factors are external to the animal, and one, biology, is innate and determines species characteristics (genetics). The directly acting factors are in turn affected by indirectly acting factors, both biotic and abiotic in nature, and through a variety of links and chains create complex interactions. The indirect factors include living organisms, inorganic matter, and energy. An understanding of the centrum and web specific to a particular animal species emerges from knowledge of ecological relationships and the growing comprehension of food chains, food webs, and nutrient cycling responsible for ecosystem functioning. The indirect factors include living organisms, inorganic matter, and energy.

While factors in the centrum account for the chance for an animal to survive and reproduce, there can be overlap between factors. For example, humans affect the survival and reproduction of an animal through hunting and habitat destruction, so humans could be classed as either or both a hazard or predator. In discovering what the components are for Proposition 1, focus is directed to an animal's Umwelt (Klopfer 1969), which can be interpreted as the totality of the surroundings that an animal experiences.

Surroundings and experiences of animals may be totally different from that perceived by humans (Figure 1.3). The challenge to forest wildlife biologists is to understand and describe the actual environment affecting wild animals and not that perceived by humans. However, it is a useful exercise to be anthropomorphic and consider which factors affect our ability to survive and reproduce. In doing so, items will be identified that can be listed under separate categories of components that directly or indirectly affect an animal's health and welfare. In most cases, indirectly acting factors are difficult to identify and quantify.

Managers spend 70–80% of their time working with only one or two of the directly acting factors in the centrum and little time on indirectly acting factors in the web. A majority of management problems occur from the directly acting factors that affect the ability of an animal to find and use the critical survival components of food and cover. However, an understanding of how the indirectly acting factors affect the directly acting factors is necessary for successful wildlife management in a forest ecosystem. Environmental and habitat terms are often used interchangeably, but they are not synonymous.

1.4.2   WHAT IS HABITAT?

The definition of habitat includes all six factors of the environment affecting the chance or ability of an animal to survive and reproduce in a specific place. One of the five directly acting environmental factors, resources, must be in an animal's habitat, but the other four may or may not be present. Factors may be present that affect one species but may not have an effect on other animals in the same place.

Habitat is the environment of and specific place where animals live: It is their home (Smith 1990; Dasmann 1981). By including environment, the definition includes all the factors affecting an animal's chance to survive and reproduce in a specific place (Patton 1992). A good discussion of the problems with the misuse of the term *habitat* by researchers, managers, and educators is contained in an article by Hall et al. (1997).

FIGURE 1.3 In the Umwelt, deer may see vegetation differently from humans. (From Jenkins, D. H., and I. H. Bartlett, *Michigan Whitetails,* Michigan Department of Conservation, Lansing, 1959.)

We generally speak of the habitat of mule deer as forest or brushland; for trout, cold running water; and for the mallard and muskrat, marshland, but all these habitats have a specific place containing environmental and biological factors. Specific places are often described by a vegetation type (oak-hickory, spruce-fir) or topographic feature (mountains, deserts). Habitat not only is for individuals but also encompasses populations and most often is thought of as a place where an animal lives without describing in detail all the components of place. When presenting information on forest wildlife, the challenge is to identify the species involved when discussing wildlife habitat. Statements like: "Timber harvesting is affecting wildlife habitat" is meaningless unless it is identified with one or more animal species that use a particular set of habitat components. Within a habitat, there may be microsites with physical features available for survival activities of feeding, nesting, calving, roosting, or hiding.

A scale factor of space and time must be considered as one part of the habitat definition. There can be *primary* habitat, which contains all the combined habitat areas and environmental factors necessary to support a viable population of a species, and there is *secondary* habitat, an area where an organism spends part of its time but all life requirements may not be present (Harris 1984). Secondary habitat seems to apply mostly to gamma or highly mobile species at the landscape scale. These animals may have to move because of weather conditions. However, this does not mean that small and medium animals (squirrels, coyotes) do not have secondary habitats, but past research has not provided information for these groups of species.

The biological factor in the centrum of an animal includes behavioral traits of migration and reproduction activities. These innate behavioral characteristics have a distinct bearing on primary habitat. For example, in the southwestern United States elk migrate from summer use of high-elevation, mixed-conifer forest habitat to a low-elevation, conifer-shrub habitat during the winter. Winter and summer habitat together must be delineated as primary elk habitat, but individually they are secondary habitats. Because there is confusion in the use of the general term "habitat," researchers have defined subsets of the term to provide a common understanding of its meaning for consistency and elimination of ambiguity (Block 1993). The following definitions were developed for ornithological use, but they provide a common understanding for discussing issues for all species.

1. **Habitat use:** The manner in which a species uses a collection of environmental components to meet life requisites. Habitat use can be regarded in a general sense or separated into specific acts or needs, such as foraging, nesting, or roosting.
2. **Habitat selection (preference):** Innate and learned behavioral responses that allow animals to distinguish among various components of the environment, resulting in the disproportional use of environmental conditions to influence survival and reproduction.
3. **Habitat suitability (quality):** The ability of the environment to provide conditions appropriate for survival, reproduction, and viability. Suitability is a continuous variable measured by the intrinsic rate of population increase.
4. **Macrohabitat:** Landscape-scale features that are correlated with the distribution and abundance of populations. Often used to describe seral stages or discrete arrays of specific vegetation types.
5. **Microhabitat:** Specific, recognizable features of the environment that act as proximal cues to elicit a settling response from an individual animal.
6. **Critical habitat:** Physical or biological features that are essential to the conservation of the species and may require special management considerations or protection.

As a result of the Forest and Rangeland Renewable Resources Planning Act (U.S. Congress 1974) and the large amount and types of data collected to make a resource assessment, it became obvious that there were several ways to categorize forest structure as wildlife habitat. Identifying habitat structures, ranging from general (macro) conditions to specific components (micro) for survival activities has led to new ways to reduce the complexity of using information in large databases.

1.5 PLANTS IN A FOREST

Plant ecology has been a formal course of study in colleges and universities for many years and is usually a required course for natural resource management professionals. The study of environmental relationships of individual plants, *autecology*, got its start in the early days of plant geographers, who were trained botanists and taxonomists. There are many good plant ecology textbooks available, and this book is not a replacement for any of those now in the marketplace. Instead, the material in this section is background information to help in understanding the relationships linking plants and animals to the same environmental factors affecting the ability of both survival and reproduction (Axiom 4). In this manner, we may discover new ways of understanding how in nature everything is directly or indirectly connected.

1.5.1 LIFE FORM

Biology is the study of life and living things, including their origin, distribution, activities, structure, and diversity, but our main interest here is the study of the biological components of a forest, specifically structure of plant *life forms*. Based on physiognomy, plant life forms in a forest can be generally described as trees, shrubs, forbs, grass, grasslike forms, moss, lichens, fungi, and so on. It would seem obvious that almost everyone familiar with the outdoors can describe a tree and its characteristics, but in scientific literature, strict plant definitions are required so that there is no misunderstanding of the item under discussion.

A *tree* is defined as a woody plant with a diameter greater than 4 in measured outside the bark at breast height (diameter at breast height, dbh) of 4.5 ft. Trees usually have a single stem with branches and leaves at some distance from the ground and are classified as broadleaf or needle leaf and deciduous or evergreen. Evergreens are those that retain some foliage throughout the year, while deciduous trees periodically lose their leaves and are therefore bare for a portion of the year.

A *shrub* is a perennial woody plant smaller than a tree, usually with many separate stems starting from or near the ground. Approximately 1,000 species of shrubs, semishrubs, and woody vines grow in the U.S. national forests and adjacent lands (Dayton 1931). Martin et al. (1951) accounted for over 100 wildlife species using shrubs for food. It is probably not an exaggeration to assert that every shrub species provides food or cover for some animal species. The association of specific animal species with specific shrub species is recognized in their respective common names, such as bearberry, rabbitbrush, deerbrush, and antelope bitterbrush. Shrubs grow in pure stands in the absence of an overstory or in scattered single or small clumps as understory plants.

Forbs are the soft-stemmed, nonwoody, wide-leafed, low flowering plants such as buttercups, cinquefoils, and clovers. Forbs can be annual, biennial, or perennial. Any plant of the Gramineae family (similar to wheat, bluegrasses, and bromes) is classified as a grass. *Grasses* have jointed stems, long narrow leaves, and usually a small, dry, one-seeded fruit. The grass family is one of the largest families of flowering plants. *Grasslike* plants look like grasses but are the sedges and rushes with pithy or hollow stems and are found in moist-to-wet soil conditions. Forbs and grasses are often grouped together as *herbaceous* plants. Herbage production follows the well-documented inverse relationship of high-percentage tree canopy/low production, low-percentage tree canopy/high production. However, total productivity is modified by soil and rainfall, so there is great variation from location to location.

1.5.2 LIFE HISTORY OF PLANTS

The complete life history of an organism covers the following time span:

- Birth
- Stages of growth to maturity
- Reproduction
- Death

Birth for plants is seed germination. Growth stages are highly variable, with the extremes of annual grasses to bristlecone pine of several thousands of years old. The maturing process produces many different plant growth forms, indicated by the changes in arrangement of stems, fruits, and leaves, to produce a recognizable shape and configuration of parts. Reproduction includes the process of flower development and fertilization, seed development and dispersal, and population dynamics resulting from plants with a few seeds to those dispersing thousands. Understanding the life history of plants is important in the process of primary succession in ecosystem restoration (Walker and del Moral 2003).

Considerable information on the life history of trees is available in Volumes 1 and 2 of *Silvics of Trees of North America* (Burns and Honkala 1990). These two volumes provide geographic range and growing conditions along with temperature and the life history factors mentioned. In addition, species are identified with other trees in their association with forest cover types of North America.

1.6 WATER IN A FOREST

Forest wildlife biologists are increasingly involved with aquatic habitats that have traditionally been the responsibility of fishery biologists or hydrologists. The information in this section supports the fact that water is a major component of forest ecosystems. The definition of wildlife includes species of amphibians, fishes, mammals, and reptiles that require water as a cover medium, a food source, or both in streams flowing through or lakes and ponds situated within forested areas.

Because management practices in forested areas directly affect water quality, it is important for forest wildlife biologists to understand some of the major biological, chemical, and physical properties of water and aquatic habitats. Excluding energy from the sun, the primary force controlling the forested ecosystem is trees; similarly, the force controlling the aquatic environment is water. Therefore, knowledge of the properties of water is a necessary requisite for management.

1.6.1 WATER PROPERTIES

The properties of water important in ecological processes are *density, viscosity,* and *transparency.* Water has its highest density at 40°F; therefore, it expands and becomes lighter above and below this point, so ice floats, allowing life to continue in the deeper water (Hickman 1966). This unique property prevents some lakes from freezing solid and thermally stratifies others. Related to density is specific heat. Water is capable of storing large quantities of heat with a relatively small rise in temperature. This property prevents wide seasonal temperature fluctuations and tends to moderate local environments. Thus, water in ponds and lakes warms slowly in the spring and cools slowly in the fall.

Viscosity is the source of resistance to objects moving through water. The faster an aquatic organism moves through water, the greater is the stress placed on the surface of an organism and the volume of water that must be displaced. Replacement of water in the space left behind by the moving animal adds drag on the body similar to the drag on an airplane moving in air. Fish, such as trout, with a short, rounded front and a rapidly tapering body meet the least resistance in water. Below the surface of the water, water molecules are strongly attracted. Above the surface, air molecules are less strongly attracted to the water; therefore, surface water molecules are drawn down into the liquid, creating a surface tension and forming a water skin. This skin can support small objects and animals such as water striders and spiders. Some mayflies and caddis flies, on the other hand, cannot escape from the skin and so become food for fish.

Transparency is the property that allows light to penetrate water. Aquatic plants, like terrestrial plants, need energy from the sun for photosynthesis (Axiom 1). Any impediment to water clearness reduces photosynthesis and therefore the amount of plant food available for aquatic organisms. The degree of clearness is termed *turbidity,* which is caused by the increase of soil particles of silt and clay or chemicals (Bennett 1986). The inorganic soil particles derived from the surrounding watershed are, however, the leading cause of turbidity.

As temperature decreases, the capacity of water to hold oxygen increases. At 32°F, water can hold about 15 ppm (parts per million) of oxygen, but this drops to 7 ppm at 104°F (Welch 1955). Oxygen can be absorbed from the air or derived from photosynthesis by aquatic plants. Warm-water fish need at least 5 ppm of oxygen, while cold-water species such as trout require amounts greater than 5 ppm (Bennett 1986). Acidity of the water affects the ability of fish to extract oxygen. A pH range between 6.0 and 8.0 seems to be optimum for the best fish production.

1.6.2 FRESHWATER HABITATS

Freshwater habitats are of two types: *lentic,* meaning calm water (lakes, ponds), and *lotic,* meaning running water (springs, streams, rivers). Both have characteristics that influence the aquatic life they contain. Based on temperature, these waters can be classified into three categories (Forbs 1961):

- Cold water, less than 65°F in summer; habitat for salmon, trout, whitefish, and grayling
- Cool water, 65–75°F; habitat for smallmouth bass, northern pike, walleye, muskellunge, sturgeon, and shad
- Warm water, more than 75°F; habitat for largemouth bass, sunfish, suckers, and carp

Fish are sensitive to temperature because it affects their oxygen supply and rate of growth. The ability to tolerate any given temperature is species dependent. For example, rainbow and brook trout can thrive in temperatures of 70°F and may be able to tolerate higher temperatures for short periods of time (Ross 1997), but they survive better in colder water (Everhart and Youngs 1981). This is probably because there is more oxygen in cold water. The ability to tolerate temperature changes seems to depend on whether the increase or decrease is rapid; slower changes allow time for acclimation.

1.6.3 LAKES

The surface temperature of lake water rises in the summer, and a warm layer, the *epilimnion*, develops. Oxygen and food are plentiful in the epilimnion as a result of the exchange between the water surface and the air and because sunlight is available for food production. This layer circulates, but it does not mix with the lower layers. Beneath the epilimnion is the *thermocline,* which is a thin layer, the temperature of which drops rather rapidly with depth (about 1°C/m). The thermocline is also defined as the plane of maximum rate of decrease in temperature (Ross 1997).

The *hypolimnion* is the lowest and coldest layer in a lake. If the thermocline is below the photosynthetic zone, then light is reduced, and the oxygen supply is depleted in the noncirculating hypolimnion. This condition most commonly occurs in summer, leading to lake stagnation. In the fall, the epilimnion cools until it is the same temperature as the hypolimnion, so circulation of oxygen and nutrients in the whole lake is restored by what is called the *fall turnover.* The turnover is caused by the mixing of the water layers by wave action. When air temperature is sufficiently cold, the temperature of the surface water drops to below 40°F, causing the water to expand, become lighter, and then freeze. If snow accumulates on the surface of the ice, light is cut off, stopping any photosynthetic processes. As a result, oxygen is depleted, and this leads to winter kill of fish. In the spring, the water warms and along with wind action creates a *spring turnover* before summer stagnation sets in again.

Thermal stratification does not occur in running water, so the plant and animal species are quite different between lentic and lotic systems. Oxygen and temperature are significant qualities in both systems, but water current is a factor only in streams, except in those lakes and ponds that are small relative to the volume flowing between inlets and outlets. But in general, lakes and ponds are closed systems. Because lakes have little, if any, water flow (closed system), *eutrophy* sets in—that is, lakes move from an unproductive state with low levels of plant nutrients (*oligotrophic*) to a state with high levels of nutrients (*eutrophic*). All lakes experience some natural eutrophication as a result of runoff containing nitrates and phosphates from their drainages but accelerated eutrophication leads to pollution when these nutrients are at high concentrations. Highly eutrophic waters are unsuitable for many aquatic species.

A new lake is oligotrophic, containing low concentrations of nutrients and plankton in the *limnetic* zone, leading to low fish populations. The limnetic zone is the zone of light penetration and consists of two areas, the *littoral* zone containing rooted plants and the *open water* zone containing floating plants. Beneath the limnetic zone is the *profundal* zone, where light does not penetrate. As the lake ages, nutrients become more plentiful, the profundal zone accumulates sediments and becomes shallower, and fish density increases due to increased vegetal growth resulting from increased nutrients and light.

1.6.4 STREAMS

The biological characteristics and development of lakes are completely different from those of running water. The current of a stream determines the physical characteristics of the stream. The velocity of a stream is controlled by gradient, width, bottom roughness, and shape (straight or meandering). Nutrients entering a stream directly from its headwaters and from adjacent areas pass through the system to the mouth, where they are deposited (open system). A multitude of microenvironments are formed within a stream system, arising out of the combination of pools and riffles produced by rocks, boulders, and bottom formation.

These combinations of physical characteristics affect the amount of oxygen and carbon dioxide in the system. In addition, the chemistry of streams is strongly influenced by the soil, vegetation, and land uses along their borders. Water percolating through the soil picks up nutrients, which in turn affect the acidity and nutrient qualities of the water. Streamside vegetation is a major determinant of water temperature by exposing the water to sun or shade. In addition, riparian vegetation provides microhabitat for aquatic insects completing part of their life cycle on land. Vannote et al. (1980) have proposed the river continuum concept by which the biological and dynamic processes of the stream are determined by the physical gradient and drainage network. Basically, the concept means that streams are small at the headwaters and get progressively wider, but in the process there is change in the kinetic energy, which in turn determines the types of organisms that are present in any given stream reach.

In the headwaters of a stream, riparian vegetation is important because it provides shading and produces detritus. In the middle reaches, algae and rooted plants provide the nutrients. As the stream widens, the effect of riparian vegetation is reduced, but there are large quantities of upstream organic matter entering the system. All the interactions between the physical characteristics and biotic components of a stream produce a state of dynamic equilibrium similar to that produced in the succession processes on land.

1.6.5 SEEPS AND SPRINGS

Water from sources that come up from the ground or from the side of a mountain that have a low volume of flow is in the seep-and-spring category. While these water sources do not contribute a significant volume of runoff, they are a rare but important source of moisture for plants such as watercress. Plants growing in moist areas tend to have a high protein content, which increases their nutritional value for many animal species. In many cases, seeps and springs are associated with a cool microclimate that attracts elk and deer in the hot summer months. Elk also use these moist areas where the soil is soft for "wallows" to mark their territory during the "rutting" season.

1.6.6 RIPARIAN ZONES

Vegetation immediately adjacent to streams or along the edges of lakes and ponds is characterized by plant species and life forms differing from those of the surrounding forest and is termed *riparian* (Figure 1.4). Riparian tree composition depends on elevation but typically consists of deciduous trees from the genera *Populus, Acer, Salix,* and *Alnus.* The marked contrasts between riparian and upland vegetation produces structural diversity and edge characteristics that enhance its utility for wildlife (Figure 1.5). Because of their long, narrow shape, stream riparian areas contribute few acres to the total available habitat; however, they are highly productive, so their value to wildlife is well out of proportion to their small area. The highest density of birds ever recorded in North America was found in riparian vegetation in Arizona (Johnson 1970). The density of 1,324 pairs per 100 ac of noncolonial birds is also among the highest densities in the world, including tropical forests. In addition to birds, riparian vegetation is habitat for beaver, river otter, and mink, along with many species of snakes and small mammals that are prey items for raptors such as the Cooper's hawk.

Not only are riparian areas important to terrestrial animals, but also they control the associated lotic habitat for amphibians and fish. Canopies provide shade, root systems stabilize banks, and plant detritus and insects provide nutrients for stream organisms (Meehan et al. 1977). Riparian areas create an oasis effect in dry lands, and because of their cooler microclimate and free water, they are major resting places

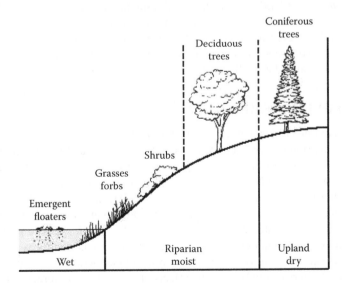

FIGURE 1.4 Riparian vegetation creates edge and diversity in a short horizontal distance.

FIGURE 1.5   High-elevation riparian vegetation along a mountain stream in Arizona.

for many north-south migrating birds. In the West, riparian vegetation provides a natural travel corridor from high mountain forests to low-elevation grasslands for species such as deer, raccoons, turkey, and bear. In the eastern United States, where higher rainfall results in more vegetation adjacent to rivers and streams, the riparian zone is not as distinct as it is in the West. The increased moisture and deeper soils in the eastern regions integrate the vegetation types, thereby lessening the distinction between where the riparian zone starts and ends.

The influence of the type extends to adjacent dryland areas, and this location is frequently designated as a riparian zone. During a symposium in Tucson, Arizona, the term *endangered habitat* was proposed, establishing the importance of riparian vegetation not only for wildlife but also for the functioning of ecosystems (Johnson and Jones 1977). Seven years later, the first North American riparian conference was held under the sponsorship of five land management agencies and the University of Arizona (Johnson and Ziebell 1985).

1.7   FOREST DIVERSITY

The National Forest Management Act (NFMA) (U.S. Congress 1976) defines *diversity* as the distribution and abundance of different plant and animal communities and species within the area covered by a land or resource management plan. A large number of species coupled with a variety of natural biological communities in an area is believed to indicate a high degree of ecological diversity. The concept of diversity in recent times has been expanded to imply that the more complex a community is, the more stable it is, and the more readily it can adjust to perturbations from within without major effects on the structure of the community as a whole. The many food chains and webs, energy pathways, and interactions in complex communities create buffers that minimize the effects of the destructive forces of nature and humankind on the community. Intuitively, diversity is thought to be "good" because complex interactions are harder to destroy than simple ones. But, the view that diversity promotes stability has been challenged (May 1973).

In a report to the U.S. Congress (Office of Technical Assessment 1987), *diversity* has been referred to as the variety and variability among living organisms and the ecological complexes in which they occur. Diversity can be defined as the number of different items and their relative frequency. For *biological diversity*, these items are organized at many levels, ranging from complete ecosystems to the chemical structures that are the molecular basis of heredity. Thus, the term encompasses different ecosystems, species, genes, and their relative abundance.

Diversity in forest ecosystems implies an increase in the diversity of habitats and consequently an increase in potential habitat to support diverse kinds of organisms (Boyce and Cost 1978). Habitat diversity to increase species richness has been used in a practical way to manage forests in the eastern region of the Forest Service (Siderits and Radtke 1977). *Species richness* is the number of species occurring in a defined area and is often referred to as *alpha diversity* (MacArthur 1965). The number of species between areas is *beta diversity*. It is a measure of heterogeneity between plant or animal communities. *Gamma diversity* includes both alpha and beta diversity in the same unit but separated from another unit by geography (Cody and Diamond 1975). In addition to alpha, beta, and gamma diversity, there are horizontal and vertical diversity. The term *species richness* is somewhat misleading because it could include those species that are rare and not common to the forest type considered. If a species only occurs in a rare instance, does that contribute to diversity or species richness?

1.7.1 HORIZONTAL AND VERTICAL DIVERSITY

Horizontal diversity results from the arrangement of plant life forms, successional stages, and varying amounts of forest components along a horizontal plane (Figure 1.6). *Vertical diversity* is the degree of divergence in height between two vegetation types or stages in a stand. It portrays the effect of layering and separation of plant life forms. Diversity created by openings and forest stands can become diverse to a point and then revert back to homogeneity over ecological time (Figure 1.7). A grass-forb successional stage adjacent to a seedling-shrub successional stage is low in contrast and thus low in diversity, but a grass-forb stage adjacent to a mature tree stage is high in contrast. In low-diversity situations, edge cannot or can hardly be detected, but as the contrast increases, the outline of the edge becomes recognizable and is fully identifiable when a high degree of contrast is reached.

1.7.2 LAYERS

A vertical line from above the tree tops to below the soil surface, at any point along a forest floor, could pass through several layers from tree canopy to underground material. Along the horizontal forest floor, the vertical strata can change from stands to openings with the contents in the space occupied by the vertical physical components.

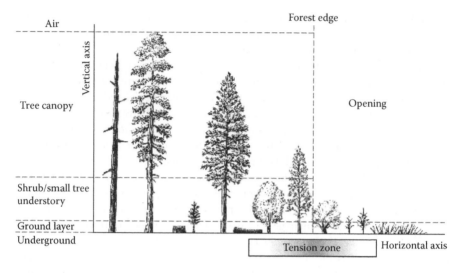

FIGURE 1.6   Horizontal and vertical diversity in a forest ecosystem.

Forest layers provide a general way to account for the vertical and horizontal diversity. These layers suggest an approach to categorize food and cover survival needs that have a level of commonality.

Management of natural resources starts with identifying and inventorying the resource base, and in the case of forest types, it is focused on the species, size, density, quality, and quantity attributes of trees. Trees are entities that have a life cycle and over their lifetime provide a significant amount of by-products in structural components. These components are contained in categories of information and are present to some degree in different combinations in all forest ecosystems. They can be translated into descriptions of natural settings, indicating what is available and identifying those that are associated with or could be used by different species of wildlife.

Forest components create structural layers (Table 1.1). These layers occur in 20 forest types in the 48 continental states. Each of the layers has sublevels (attributes) of information that refine data to associate an animal species with forest components. For example, *volume* contains the size, age class (i.e., mature), and density (i.e., thickness) of trees in a combined metric. In some common forestry and wildlife usages, tree size, age, and density are referred to as *stand structural stages*.

1.7.3 COMPONENTS

As a result of the continuing effects of federal legislation since 1976, managers continue to need new and different ways of associating animal species with their environment and habitat at different levels of detail for management purposes. One way to do this is to consider all the *biotic* and *abiotic* components that constitute a forest and then organize these components in a way that they can be identified with species of wildlife that use them for food, cover, and biological activities. In a review of publications relating to forestry and wildlife, four categories of terms were

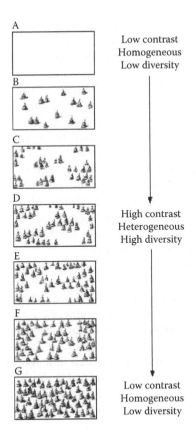

A

Low contrast
Homogeneous
Low diversity

B

C

D

High contrast
Heterogeneous
High diversity

E

F

G

Low contrast
Homogeneous
Low diversity

FIGURE 1.7 A visual representation of diversity going from openings to forest stands.

TABLE 1.1
Structural Components of a Forest

Layer	Descriptor
Air	Space
Canopy	Overstory
Volume	Stand size, density, and layers
Opening	Less than 10% canopy
Shrubs/sapling	Understory
Ground	Forest floor
Underground	Tunnels, burrows, and the like
Water	Water in a forest environment
Microsites	Nests, beds, burrows, and the like

most commonly identified that were related to forest situations. The categories listed below are not all inclusive but represent the major components that a field biologist would encounter in evaluating forestry and wildlife projects.

- **Physical components:** Boulders, talus, air, duff-humus, woody debris, soil, leaf litter, logs, stumps, snags, gravel, caves, amphibians, birds, fish, tree crowns, mammals, reptiles, invertebrates, trees, shrubs, grass, seeps-springs, water, fungi, tree trunks, tree branches, rocks, lichens, moss, vines, aquatic plants, ground floors, tree bark, tree trunk
- **Synthetic descriptors:** Cover type, tree size, tree density, riparian, over-story, understory, coniferous, deciduous, even aged, uneven aged, above ground, below ground, opening, tree stand, old growth, ground surface, mature, young, intermediate, single story, two story, multistory, canopy
- **Physical microsites:** Nest, bed, roost, perch, lodge, den, cavity, wallow, burrow, tunnel, cliff-ledge, platform
- **Biological microsites:** Breeding, hiding, resting, thermal, travel, escape

The structural layers identified in Table 1.1, when combined with factors in the four categories described, can be rearranged for efficient use in a data model. Table 1.2 contains attributes of descriptive information in a menu to select items for food and cover areas that are used by an individual species of wildlife. The information contains codes arranged so that they can be used in a systematic way in a relational database. Table 1.2 is a core table designated as *Life-Needs* in a data model, FAAWN (forest attributes and wildlife needs), which is discussed in Chapter 5 and is included on a compact disk attached to the back cover of this book.

1.7.4 ARRANGEMENT OF COMPONENTS

The minimum arrangement of food and cover in close proximity to one another for a species is termed *juxtaposition* (Bailey 1984). Food available in one area (opening, meadow, etc.) and cover (tree stand, shrubs, etc.) nearby clearly reduces the energy required for and the natural hazards associated with the search for food away from cover. Equally clear is that the expenditure of energy to search for food and cover cannot be greater than that obtained from the food source, or the animal loses weight. Another term related to juxtaposition is *interspersion,* which is the distribution of habitat units for a species over the landscape. The arrangement of stands to provide food and cover in a nonsystematic way provides a measure of diversity in ecosystems and landscapes to meet the habitat requirements for a large number of species.

1.7.5 BIOLOGICAL DIVERSITY

The term *diversity* has been in use by wildlife biologists in indirect ways for many years. Leopold (1933) talked about the desirability of simultaneous access to more than one environmental type or the greater richness of border vegetation. Pimlot

TABLE 1.2
Attributes Menu of Basic Forest Components Formatted for Use in the Life-Needs Table in the Relational Data Model in Chapter 5

Cname. Common name

Lform. Life form

Acode. Animal name code

SuRe. Survival and reproduction factors

    Fo = food area

    Cover = cover area

Air. Space for feeding

    a = feeds in air over/through canopy

    b = hunts from air (swoops, hawks)

    c = feeds while hovering

Ov-story. Overstory stand canopy factors

    a = upper canopy

    b = general canopy/crown

    c = low canopy

        1 = dead tree/snag

        2 = trunk

        3 = bark

        4 = isolated wolf tree

        5 = edge (forest, openings, stages)

Volume. Stand size, density, and layers

    sa = sapling (1- to 4-in. dbh)

    yo = young (4–10 in.)

    in = intermediate (11–17 in.)

    ma = mature (>17 in.)

    ol = old forest (mature + conditions)

        a = single story

        b = two story

        c = multistory

            1 = thin (10–40%)

            2 = thick (41–70%)

            3 = closed (>70%)

            + = increase value

Un-story. Understory vegetation

    a = shrubs/saplings

    b = herbaceous (grass/forb)

Opening (<10% canopy)

    a = grass/forb

    b = shrub/sapling

    c = wet meadow

Ground. Ground material

    a = logs/stumps

    b = duff-humus

    c = boulders

    d = rock field/talus/crevices

    e = elevated cliff/ledge

    f = surface debris

    g = underground

Water. Water in a forest environment

    a = riparian

    b = near water

    c = water medium

    d = aquatic vegetation

Microsites. Specific food and cover areas

    1 = tree/shrub nest

    2 = ground nest/bed

    3 = tree cavity

    4 = ground cavity

    5 = in/on water

    6 = tree perch, roost

    7 = platform

Inform. Additional information

(1969) provided a detailed review of the topic, and Dasmann (1981) introduced the idea of "declining diversity" in his *Wildlife Biology* textbook. The current definition of biological diversity has evolved to include the variation of life at all levels of organization (Gaston and Spicer 2004).

Because *biological diversity* is a complex subject, it is useful to consider the different levels from the narrowest to the broadest perspective (U.S. Department of

Agriculture [USDA] 1990). Four general categories (genetic, species, ecosystem, and landscape) illustrate the range of complexities contained in the term:

- *Genetic diversity* refers to inherited characteristics found among individual representatives of a species that are critical to the survival of the species. A variety of genes provides for resilience under environmental stress, which allows a species to adapt to changing conditions.
- *Species diversity* is the level of biodiversity that receives the most attention. It is the variety of living organisms found in a particular place, for example, the hundreds of different species inhabiting a ponderosa pine forest, including all life forms of plants and animals. To address species diversity, one must consider how the species changes from place to place as well as over time in the same place.
- *Ecosystem diversity* includes the variety of species and ecological processes (nutrient cycling and energy flow) that occurs in different physical settings, such as in an old-growth forest, a riparian zone, or a desert habitat.
- *Landscape diversity* (gamma diversity) deals with the dispersion of ecosystems, such as grasslands, wetlands, ponds, streams, and forests, in a geographical setting. A broad expanse of forests or grasslands is not as diverse as the same expanse containing a mixture of ecosystems.

Conservation biology is a discipline that has resulted from understanding the four biodiversity categories with the emphasis on the protection of species, their habitats, and ecosystems (Soule 1986). A somewhat broader definition is:

A science that conducts research on biological diversity, identifies threats to biological diversity, and plays an active role in the preservation of biological diversity. (Primack 2006, p. 530)

Implicit in these definitions is rates of extinction, which ultimately encompass population demographics, viable populations, and threatened and endangered species.

Biodiversity and conservation biology were presented in a clear and different way as related to forests by Lindenmayer and Franklin (2002, p. xi). Their approach was to examine the "matrix" outside the "set-aside" or reserved lands and how "management practices on these lands will largely determine how successful human society is at conserving forest biodiversity and maintaining forest health." The five chapters in this textbook consider how components in forest types, in the matrix, at multiple scales, contribute to the food and cover need of over 1,000 forest animals in 20 forest types in 48 states.

One of the disciplines supporting biodiversity and conservation biology is landscape ecology (Turner et al. 2001; Bissonette 1997). The need exists for a landscape-level analysis developed from the ability to understand space as it relates to size, distribution, and juxtaposition of communities and ecosystems. As humans, we see large areas in our immediate field of vision and call it a landscape, but landscapes, like ecosystems, have no definite space limitation. Of significance is the arrangement of landscape components into recognizable patterns and how these patterns are distributed now, over time, and what causes these patterns to change.

The need for a landscape view has resulted from having to manage large areas following natural disturbances such as fire, floods, and hurricanes or human-induced changes such as timber harvesting. All of these factors start a process that results in different plant and animal communities. Where these events occur, there may be need for a program to rehabilitate the land to a previous or different vegetative state (Walker and del Moral 2003). As with any new discipline, there is also new terminology, such as matrix, patch, and metapopulation, and these terms are now common in ecology journals and textbooks (Bissonette 1997; Lindenmayer and Franklin 2002; Smith and Smith 2009; Millspaugh and Thompson 2009). Techniques that are important for collecting landscape data include remote sensing, aerial photography, and geographic information systems, all of which require some level of on-the-ground verification. Aided by powerful computer storage and retrieval software, landscape analysis is becoming a necessary management tool for natural resource managers.

As with any new way of doing business, software is often developed that facilitates the use of large amounts of data for collation and analysis. This has happened in landscape ecology, with many ways to collect information on forest components, including wildlife populations (Morrison 2002; Skalski and Robson 1992), along with the earlier developments for using spatial statistics (Ripley 1981) and planning models (Millspaugh and Thompson 2009). With all the new information that is now available in the literature, natural resource managers may be facing the paradox of choice: More is less (Schwartz 2004) and "good enough" may be the best that can be expected or attained. In this textbook, *choice* refers to goods and services, and I interpret information as a service.

1.7.6 EDGE

When two distinct vegetative life forms (trees, shrubs, grass) or structures within life forms (mature trees, saplings) meet, a boundary or *edge* is created between the two (Figure 1.6). While tree stands and openings are major structural components, there is an area of transition between the two that is a significant part of the forest. The blending of trees into an opening, openings into a less-dense tree area, and young stands into mature stands creates a transition zone with a variety of habitats with more opportunities for food and cover for different species of animals. This change from one life form to another or from one vegetative structure to another intrinsically produces an increase in variety or diversity. Leopold (1933) referred to this change or increase in diversity as "edge effect." Leopold viewed animal habitat as a phenomenon of an edge, from which he developed the law of dispersion, which states that the potential density of game of low radius requiring two or more types is, within ordinary limits, proportional to the sum of type peripheries.

Leopold's law is based on the premise that home range is fixed for a species as an innate characteristic and determines how far an animal will go in search of food and cover. Although Leopold's law of dispersion was developed in relation to small game and nongame species, his edge principle applies in varying degrees to all wildlife (Figure 1.8). Efforts to understand edge and how it can be used in wildlife

FIGURE 1.8   Elk at the edge of a wet meadow in a fir-spruce forest (Type 26) on the Apache National Forest in Arizona.

management have evolved into more profound meanings and techniques for quantification of the concept. Edge can be of two types: inherent and induced (Thomas et al. 1979). An *inherent edge* is the natural outcome of the evolution of successional and structural stages of plant communities. *Tension zones* arise when the position of the naturally evolved inherent edge is in constant change (Figure 1.6).

Induced edges result from management practices such as fire, timber harvesting, grazing, or seeding and planting. All are direct but short-term activities undertaken to manipulate succession. Both types of edges can be either abrupt or meandering, depending on the nature of the vegetation types or successional stages. The creation of edges has been one of the major tools used by wildlife managers to increase game species. While many game species, such as the white-tailed deer, benefit from edge, nongame species or species needing large blocks of land with interior habitats often do not. The wildlife manager's adage (Axiom 4) that what is done to benefit one species will in turn be detrimental to another still holds.

1.8   VEGETATION

Vegetation in the most general sense is the aggregate of all plant species in a particular area without defining its floristic composition, physical appearance, or size of the occupied area. By observing vegetation along a horizontal and vertical plain, we can easily see variations in plant life form and plant density, which together give an impression of vegetation structure or *physiognomy*. The number of plants per unit area creates *density*, which is a specific quantitative measure, such as number of trees or shrubs per acre.

When considering vegetation over a large area, we tend to think in subjective terms of abundance. *Abundance* indicates the visual characteristic, thick to thin or open and closed, without providing a quantifying number, but it does relate a sense of compactness. *Frequency* is a measurement to describe how often a plant occurs. Quantitatively, it can be described as a ratio of number per unit of measurement. The unit of measurement can be an area or count of plants. One species of tree found on 10 of 100 plots has a frequency of 10%, or 30 oak of 100 trees counted is a frequency of 30%. A close look will reveal that vegetation is not homogenously spaced unless it is a plantation developed by humans. Plants can be randomly spaced, or they can be in small groups or even small groups that are randomly spaced. The effect of grouping and dispersal is referred to as *sociability*.

Plant life form, density, and spacing in vegetation create vertical layers. This effect is most prominent in a forest, where there is tree overstory and a shrub and grass understory. The trees themselves can create several layers based on dominance characteristics of the species. *Dominance* generally refers to the most abundant or most prominent species but in the case of trees can also indicate tree height in relation to how much light it is receiving through the canopy. A *dominant* tree receives the most sunlight from the top and sides; a *codominant* tree receives light mostly from the top, and an *intermediate* tree receives less than the full amount. *Suppressed* trees receive little sunlight.

1.8.1 Classification

In ecological literature, different terms (i.e., biome, biotic community, life zone, plant community, formation, association, ecosystem, cover type, vegetation type, and habitat type) are commonly used to describe vegetation at various levels of integration. In the past, there has been considerable effort by ecologists to delineate vegetation over continental or large regional areas, for example, biomes (Shelford 1963); life zones (Merriam 1898); biotic provinces (Dice 1943); forest and range ecosystems (Garrison et al. 1977); North American biotic communities (Brown 1998); North American terrestrial vegetation (Barbour and Billings 2000); and North American forests (Walker 1999).

Beginning in the mid-1980s, there was an effort to develop a classification that would bring some commonality to naming vegetation and plant communities so that ecologists would speak the same language. *Classification* is the grouping of things with similar characteristics and reduces heterogeneity depending on the scale or level of abstraction of the groupings. A major effort by the United Nations (1973) resulted in a worldwide vegetation classification scheme. The system is open ended and hierarchical, using plant physiognomic characteristics for descriptors. Under the UNESCO (United Nations Education, Scientific, and Cultural Organization) system, there are five mutually exclusive vegetation classes in the first level of the hierarchy:

    I. Forestland
    II. Woodland
   III. Shrubland
   IV. Dwarf shrubland
    V. Herbaceous vegetation

As habitat for forest wildlife, we are interested primarily in Classes I (Forestland) and II (Woodland) because they contain trees as the dominant plant life form. The distinction between the two classes is the degree of canopy closure. The woodland class has a more open canopy, in the range of 25–60%. The UNESCO classification is further subdivided into subclass, group, formation, series, and association/habitat type. At the association level, the first name is the apparent dominant or codominant plant species in the tallest life form. The second name may be a dominant species in a different layer, with a third name added if there are three layers with a conspicuous plant component, such as

Class	Forestland
Subclass	Evergreen forestland
Group	Needle leaved
Formation	Rounded crowns
Series	Ponderosa pine
Association	Ponderosa pine/antelope bitterbrush

Vegetation over large landscapes is not too difficult to identify to a low level of continuity or commonality. In mountain areas of the western United States, even the most casual observer can see that as elevation increases from 1,000 ft to several thousand feet, vegetation changes, and some species of plants will become common and dominate the scene. Observers most likely will attribute this change in plant species to temperature and moisture differences because they also can physically experience the same elements. In addition, observers will notice that vegetation along the route changes gradually, and at close range a boundary is not distinct. They will notice that vegetation is not entirely homogeneous, but there are locations where plants form areas containing different species that seem to be influenced by local conditions, including soil and exposure to sun on a slope.

The effect of climate and change in elevation creates a distinct and identifiable vegetation zone. Such a zone was recognized by Merriam (1898) in his travels in the western United States as a member of the U.S. Biological Survey. Merriam identified his life zones with increasing elevation as Sonoran, Transition, Canadian, Hudsonian, and Alpine. These life zones are useful in the West in mountainous country, but the concept is difficult to apply to lower elevations across the United States. Merriam's life zone classification was indicative of vegetation descriptions plant ecologists were trying to formulate to create a common understanding of classification that would allow discussion about areas where vegetation exists, why, and what causes it to change. Barbour and Billings (2000) listed 17 formations for North America.

The names of the formations consist of combinations of life forms, climate, geography, and plants. A chapter for each formation in Barbour and Billings' publication describes its ecological characteristics in detail and is an excellent reference for students and researchers. A subunit of a formation can be a vegetation type named by its dominant species, such as ponderosa pine, oak-hickory, beech-birch-maple, or redwoods. Within the continental limits of the United States, forests have been defined and delineated into major types containing native tree species. Forest type has been a classification system in use by the USDA Forest Service since 1929 and is widely accepted by land managers, plant ecologists, and wildlife biologists (Eyre 1980). Tree diversity in the types results from the interactions of the range of soil, topography, and climatic conditions found within the geographical areas for each type.

In the literature, the use of cover is both a forestry and a wildlife term, but with different meanings. *Cover* or cover type in forestry and plant ecology is the vegetation that overlays soil material. In wildlife, *cover* is used to designate areas for protection and shelter of animals. In this textbook, the term *cover type* is changed to *forest type* to avoid confusion between the two definitions. The criteria used to describe forest types are as follows:

- The dominant trees must cover at least 25% of the area.
- The type must occupy a fairly large area but not necessarily in continuous stands.
- The type must be based entirely on biological considerations.

One of the first maps to depict forest vegetation by regions in the United States was published in a geography textbook (Tarbell 1899). Basically, the map delineated the country into two forest regions: coniferous and deciduous. The northern part of the United States extending along the border with Canada was coniferous, the middle part from Ohio south to Tennessee was deciduous, and the southern portion from Louisiana to Florida was coniferous. All of the forested area west of the Great Plains was shown as coniferous except for the middle and southern part of California, which was designated as deciduous.

Shantz and Zon (1924) published *Forest Vegetation of the United States in Natural Vegetation: Atlas of American Agriculture*. The 1924 map was also delineated into western forest vegetation and eastern forest vegetation, but under these two designations there were specific vegetation types (nine in the west and nine in the east). The Shantz-Zon map was interpreted and reformatted by the U.S. Forest Service to create *Forest Regions of the United States* (USDA 1948). Later, Kuchler (1966) developed a map of potential natural vegetation that was also included in the National Atlas. His vegetation map is widely used because it includes shrub and grassland, not just forestland and forest types. A later version of the map was produced by the U.S. Forest Service (USDA 1967) to be included in the U.S. National Atlas and has continued to be used to depict the distribution of forest vegetation across the United States. The 1967 map, along with Kuchler's 1966 map, was interpreted for a map, *Forest and Range Ecosystems of the United States* (USDA 1974).

The most recent work on forest types was produced by advanced very-high-resolution radiometer (AVHRR) composite images during the 1991 growing season (Zhu and Evans 1994) and is included in the National Atlas (USDA 2000). All of the forest types are basically the same as when first described on the 1924 map by Shantz and Zon. The latest National Atlas map validates the excellent fieldwork to document type location and ecological descriptions by early plant ecologists.

1.8.2 Stands of Trees

A forest is composed of trees in various combinations of size and density and, depending on these two attributes, can form different layers of foliage and trunk volume in groups of trees. To document how animals survive and reproduce in a forest environment, we need to identify what life form, natural factors, and qualitative and quantitative attributes are available for use as food and cover. One way to do this is to consider a stand of trees as the basic descriptor of forest structure.

> A stand is a community of trees possessing sufficient uniformity in age, spatial arrangement, or conditions, to be distinguishable from adjacent communities and forming a silvicultural or management entity. (Ford-Robertson 1983, p. 254)

A stand does not have to be even aged but can be homogeneous units of uneven-aged trees. The definition presented for trees is often extended to herbaceous vegetation as well, such as a stand of grass, forbs, or shrubs. Stands of trees are small ecosystems in a forest that can be managed as parts of larger units across a landscape.

1.8.3 Openings in Stands

Stands are the basic unit for managing forest ecosystems, but as indicated, forests contain components other than stands. Open areas without trees are part of the forest, as are natural meadows or rocky areas with little or no vegetation. There is no general definition of an opening, but for descriptive purposes it is considered as a column of space in a forest with less than 10% tree canopy in overstory trees.

An opening is not a solid physical feature, but it is considered a structural stage because the space above the shrub/small-tree layer is space used by many birds as a zone for feeding from the air or hunting from perches in scattered trees.

1.8.4 Dead Trees and Snags

While a live tree provides food, homes, and special use sites for many animal species, use continues after the tree dies. Just as a live tree goes through developmental stages, a dying tree goes through decaying stages, from having a solid trunk, with brown needles or leaves, to the final stage during which most of the bark and branches are gone, and the trunk has become soft and easy to penetrate. Large amounts of dead and decaying woody material produced by trees can be a serious fire hazard to property, forest animals, and human life. As a result, there is a trade-off in how much material should be retained versus the risks of retaining fire hazards, especially in a wildland urban interface. However, dead trees are as important to ecological functioning in a forest as live trees (Franklin and Forman 1987), and this also has to be taken into account.

As a by-product of live trees, dead trees become snags through the decaying process. Dead trees provide different types of wildlife homes and food than when they were alive and healthy. After the snag has fallen to become a decaying log, it is continually reduced to organic matter and its nutrients returned to the soil to be used again, such as by a seed when it falls from a tree to become part of a seed bank, sprouts, roots, and starts to grow. Dead trees, snags, hollow trees, decaying logs, and stumps are used as hunting perches for raptors; den trees for raccoons, squirrels, and bears; feeding sites for woodpeckers; and nests for cavity-nesting birds. On occasion, rabbits use canals, resulting from decayed roots, for tunnels and dens, as do snakes, lizards, and small rodents. Cavity-nesting birds that use snags are a particularly important class of forest birds because the majority are insectivorous and help control endemic forest insects that damage valuable timber trees. Recognition that snags and dead or dying trees are important components of the forest is not new (Axiom 8).

Richard Jefferies (1879) provided the reasons that a hollow dead tree should not become a candidate for the axe. He eloquently described a dead and decaying tree as an ecosystem:

> Those hollow trees, according to woodcraft, ought to come down by the axe without further loss of time. Yet it is fortunate that we are not all of us, even in this prosaic age, imbued with the stern utilitarian spirit; for a decaying tree is perhaps more interesting than one in full vigour of growth. The starlings make their nests in the upper knot-holes; or, lower down, the owl feeds her young. ... The woodpecker comes for the insects that flourish on the dying giant; so does the curious little tree-climber, running up the trunk like a mouse; and in winter, when insect-life is scarce, it is amusing to watch there the busy tomtit. ... For the acorns the old oak still yields some rooks, pigeons, and stately pheasants, with their glossy feathers shining in the autumn sun. Thrushes carry wild hedge-fruit up on the broad platform formed by the trunk where the great limbs divide, and, pecking it to pieces, leave the seeds. ... Sometimes wild bees take up their residence in the hollow, slowly filling it with comb, buzzing busily to and fro; and then it is not to be approached so carelessly, though so ready are all creatures to acknowledge kindness that ere now I have even made friends with the inhabitants of a wasp's nest. ... In the evening the goat-sucker or nightjar comes with a whirling phantom-like flight, wheeling round and round: a strange bird, which will roost all day on a rail, blinking or sleeping in the daylight, and seeming to prefer a rail or branch without leaves to one that affords cover. Here also the smaller bats flit in the twilight, and, if you stand still, will pursue their prey close to your head, wheeling about it so that you may knock them down with your hand if you wish. ... All the year through the hollow tree is haunted by every kind of living creature, and therefore let us hope it may yet be permitted to linger awhile safe from the axe. (pp. 79–81)

While Jefferies' description relates to snags, it is axiomatic that the same process and use by wildlife can be extended to coarse woody debris.

1.8.5 Coarse Woody Debris

Woody debris is generally considered as logs, stumps, and limb material with a diameter over 3 in (Harmon et al. 1986; Brown et al. 2003). This type of dead material will go through a decaying process and return nutrients to the system both on land and in water (Harvey et al. 1987). During the decomposition process on land, the woody material becomes microhabitat for insects, snakes, small rodents, rabbits, and ground squirrels. In mountain streams with riparian vegetation, debris can reduce water flow to create pools, provide nutrients to the aquatic system, and increase insect populations as a food source for fish.

Woody material adds another diversity component to the already-complex forest structure. It is a component that can be beneficial for several species of chipmunks, ground squirrels, rabbits, and ground-nesting birds. Coarse woody debris can provide hiding cover for large animals such as deer and elk by reducing their visibility along roads and trails and in large openings. Logs and stumps are forest components that occur either from natural processes or left from logging operations. Because of their size, logs and stumps offer opportunities for use by skunks, fox, and rabbits. While there has been considerable research on the use of snags by cavity-nesting birds, there has not been a corresponding amount of research into the use of logs and stumps for cover, especially in a quantifiable way, for land use planning.

1.9 FORESTS AS ECOLOGICAL SYSTEMS

Scientists, in trying to describe and understand the complicated world in which we live, are constantly developing new ways to classify the enormous number of diverse plant and animal species into recognizable categories of like things for practical understanding and awareness. The method that has evolved to comprehend living things is a hierarchical system that progresses from individual organisms, through populations and communities, to ecosystems. Classification schemes are a product of thinking in an organized manner, which tends to be hierarchical. Hierarchies are a convenient method of abstraction that allows us to reduce complex systems to simpler units we can understand and study (O'Neill et al. 1986). But, information at each level of the hierarchy can be aggregated upward. This same procedure can be used regardless of the type of data contained in each distinct level. Hierarchical descriptions also provide a way for managers and researchers to communicate. Thinking in hierarchical terms allows us to discover how individual levels are integrated into more complex systems.

1.9.1 Hierarchical Organization

One way to visualize the differences in the organizational levels in a hierarchy is to progressively add to the complexity, starting with a single individual as the first level of integration or resolution. *Individuals* are distinct and cannot be subdivided into units that can live as single entities. An individual is uniform in its genetic composition, but there are differences between individuals in a population. As more organisms of the same species come together in an interbreeding group, the result is a *population*.

A population of one species becomes a *community* when added to a population of another, and the various combinations of plant and animal communities create a degree of ecological diversity. Therefore, we speak of several populations of different plant species as a plant community or of animals as an animal community. Animal communities cannot exist without plant communities, and plant communities cannot exist without nutrients and energy (Axioms 1 and 2). The addition of these two factors to the combination of existing plant and animal communities is an *ecosystem.* The term *ecosystem* was early defined as the whole complex of physical factors forming the environment (Tansley 1935). Odum's (1971) definition clarified the concept:

> Any unit that includes all the organisms in a given area interacting with the physical environment so that a flow of energy leads to a clearly defined trophic structure, biotic diversity, and material cycles is an ecological system or ecosystem. (p. 8)

1.9.2 STRUCTURE AND PROCESSES

Basically an ecosystem is an energy-nutrient processing system. Ecosystems have physical *structure,* as for example soil, plants, and animals, as well as *processes* of energy flow and nutrient cycling involving biotic and abiotic components. In addition, ecosystems have time and space attributes. While the term *ecosystem* has been used in the scientific literature since the 1930s, there is often reluctance by managers to embrace the concept because they do not directly consider energy and nutrients in the decision-making process.

In the hierarchical system of organization, an ecosystem describes an area and the animals and plants (*biotic components*) within it as well as their abiotic environment (*physical components*). In all cases, the ecosystem functions as a unit through nutrient cycling and energy flow through plants and animals (Axiom 6).

Nutrient cycling differs from energy flow as nutrients eventually return to the system through the death and decay of plants and animals in which they are temporarily bound. Functional structure develops from trophic levels (often called *biotic pyramids*) from the energy flowing from a large number of producers (green plants) to a smaller number of primary and secondary consumers and ultimately to decomposers.

One example of a biotic pyramid computed for biomass in a forested area is for ponderosa pine in Arizona (Figure 1.9). Aboveground biomass is approximately 51,000–73,000 lb/ac for producers and 7–11 lb/ac for consumers (Clary 1978). The producers include ponderosa pine and Gambel oak trees, woody shrubs, and herbaceous plants. Animal consumers include cattle, elk, deer, small mammals, birds, and insects. Practically all of the biomass of producers (96%) was accounted for by the woody by-products of trees; the remaining 4% of the total weight was in herbaceous plants and foliage. The consumer biomass is only 0.01–0.02% of the total weight.

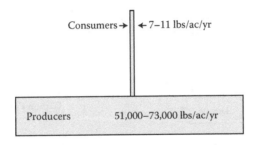

FIGURE 1.9   Producer-consumer biomass in a ponderosa pine forest in Arizona.

1.9.3 NUTRIENT CYCLES

All elements go through a cycling process from their reservoirs of living organisms, air, water, and soil. Three types of cycles exist: *biological, geological,* and *chemical.* Collectively, these single cycles are often referred to as *biogeochemical* cycles. While about half of the basic elements are essential for life, four (carbon, nitrogen, oxygen, and hydrogen) constitute about 97% of the weight of protoplasm (Emmel 1973) and cycle most easily (Figure 1.10). Carbon is a basic building block of the molecules necessary for life, so that the cycling of carbon through the use of CO_2 (carbon dioxide) by plants and respiration by animals is a critical life-sustaining process.

At death microbes (*decomposers*) break down the organic matter in a decaying process that releases carbon bound in animals to the soil, air, and water reservoirs. There is a growing concern that carbon dioxide is increasing because of increased human activities. Living animals take in O_2 (oxygen) and release CO_2 from respiration. As a result, CO_2 is contributing to the greenhouse effect, and the theory is that this is causing an increase in global warming. The clearing of forests for agriculture can decrease carbon held in the trees as much as 90% (Houghton 1995). In the Pacific Northwest, there is a question whether forests are a source or sink for carbon (Cohen et al. 1996). Global warming and the CO_2 question will remain for a long time.

In the southwestern United States, there exists a good example of nutrient cycling in a food chain of the tassel-eared squirrel, which returns nitrogen from feeding debris back to the soil for use by ponderosa pine trees. This tree squirrel feeds on pinecones, twigs, and hypogeous fungi growing beneath the duff layer of pine stands. Research has substantiated that approximately 5.32 lb/ac/yr more nitrogen is returned to the forest floor from squirrels feeding in pine trees in comparison to trees where there is no feeding activity (Skinner and Klemmedson 1978). Other research has documented that squirrels feed on mycorrhizal fungi that are obligate symbionts of the pine root system (States 1979). By feeding on the fungi, the squirrels, through their

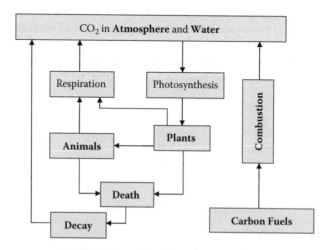

FIGURE 1.10 Carbon cycle in an ecosystem.

fecal material, spread spores of the fungi that are root inoculums for bacteria. The bacteria produce mycelium, which facilitates the cycling of nutrients from the soil to the root system of trees. This same association of fungi and rodents has been documented for other areas, such as the Pacific Northwest (Trappe and Maser 1978).

1.9.4 ENERGY FLOW

About 67% of the total energy of the sun warms the atmosphere and land surface at noon on a clear day (Gates 1965). It takes approximately 8 min for light energy to arrive on the surface of the earth, and less than 2% is trapped by plants and stored as chemical energy (Emmel 1973). The primal source of energy for all of the biological systems of the earth is the sun (Figure 1.11). The air we breathe is composed of 78% nitrogen, 21% O_2, and about 0.03% CO_2. The chlorophyll in the leaves of green plants absorbs a small portion of the energy of the sun, which is then used to combine CO_2 from the air and water (H_2O) from the soil into starches and sugars, which are contained in the plants eaten by animals. Thus, plants (autotrophic) depend on the energy of the sun, chlorophyll, and photosynthesis for their existence (Axiom 1); in contrast, *animals do not produce their own food*. It is important to remember this fact when reviewing theories about habitat, wildlife populations, and factors regulating populations.

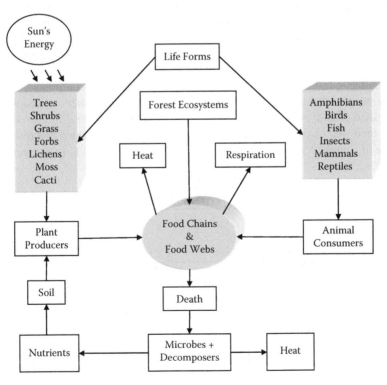

FIGURE 1.11   Energy flow in an ecosystem.

Energy flow is a one way transfer from plants to animals in an herbivore relationship or from animals to animals in a carnivore relationship. Energy has been changed in the transfer, but none has been created or destroyed; it has merely changed in form (*first law of thermodynamics*). During the transfer process, a loss of energy as heat through respiration always occurs, so to keep the ecosystem functioning there must be a constant flow of energy from the sun through the system (*second law of thermodynamics*).

1.9.5 NUTRIENT AND ENERGY PATHWAYS

Energy flow and nutrient cycling in an ecosystem requires pathways. These pathways are found in animal *food chains* and *food webs* (Axiom 6) (Figure 1.12). In simple ecosystems, as in the arctic, the pathways are short and few, but in many ecosystems, such as the tropics, the pathways are long, numerous, and hence complex. Whatever the length and complexity, every food chain must start with plants (Axiom 2). Energy from plants is transferred through a series of animal species, each of which eats one or more species preceding it in the chain. The shorter the food chain, the greater the energy available for converting food items into living matter and metabolic activity. Food chains and food webs are the *engines of ecosystems,* and without them there would not be sustainable life. Anytime a chain is broken, a link in the system is eliminated. The question is, How many chains can be broken before energy and nutrient paths are degraded to a point at which life cannot be sustained for an individual species?

In any food chain, *waste matter* applies only to an individual organism since waste such as fecal material from deer provides food and nutrition for other organisms (dung beetles, soil organisms, plants, etc.). Actually, there is no waste matter, only artifacts. In the case of humans, some of those artifacts are plastics that may take hundreds of years to decompose into elements usable by other organisms. When many food chains exist in an ecosystem, an interlocking of the pathways develops into a food web (Axiom 6). As more animal components are added to a food web, a degree of complexity is introduced. This complexity is thought to induce ecosystem

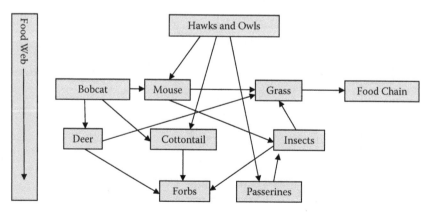

FIGURE 1.12   Food chains and food webs are the engines of an ecosystem.

and population stability, so that the loss of one food species results in adjustments by animals to another food source (Hutchinson 1959). The idea of stability is often linked to complexity, but mathematical analysis has shown this may not be the case (Goodman 1975; May 1973). The hypothesis is that the more complex a food web is, the more stable the populations in the web are. One of the problems has been the way stability is defined, and this remains a matter of dispute.

One definition of *stability* refers to an ecosystem in which the chance of a species going to extinction is low (Colinvaux 1986). For an ecosystem (or population) to be stable, it must demonstrate *resilience*, that is, the power, when under stress, to return to its original state when the stress is removed (Holling 1973). Theoretical ecologists discuss stability in terms of the number of species, and if this number remains the same over time it is said to be *persistent* (Lewin 1986). On the other hand, a fluctuating density of animals within a species refers to *constancy* (fluctuating numbers indicate low constancy; steady numbers indicate high constancy). However, the definitions and application of stability, resilience, persistence, and constancy are all theoretical and remain an issue of debate.

1.9.6 SOIL

While resources directly affect the chance of an animal to survive and reproduce, soil nutrients directly affect the ability of plants to survive and reproduce (Axiom 1). Therefore, soil is one of the indirectly acting factors affecting the welfare of an animal. It is well documented that agricultural production is directly related to the level of nutrients in the soil.

Nutrients in the soil are transferred to nutrients in plants, which are in turn transferred to animals when consumed. If the nutrients in plants are sufficient in quantity and proper relationships, the consuming animal remains healthy (Axiom 2). This is as true for wildlife as it is for domesticated animals. Studies in Missouri have shown that the highest density of game animals (rabbits, raccoons, etc.) is in areas of the highest soil fertility (Crawford 1950).

When mule deer are removed from low-fertility California chaparral and fed optimum diets in confinement, they reach a larger size after the first year of feeding than some adults in the wild attain in a lifetime (Dasmann 1981). Results of research in Pennsylvania on white-tailed deer followed the same trend as in California (French et al. 1956). Other research has shown that the number of young produced by cottontails (Hill 1972) and white-tailed deer (Cheatum and Severinghaus 1950) is correlated to increasing soil fertility. In other studies on white-tailed deer, body weight has been shown to be related to soil fertility (Smith et al. 1975).

Soil is a major factor influencing vegetation, and vegetation is the driving force that translates into food and cover for wildlife habitat (Axiom 2). Perhaps the most important soil and wildlife relationship is that wildlife production, like any other crop of the land, is related to fertility of the soil. Allen (1974) emphasized that in terms of fertility, soil and water are equivalent since water is needed to make the soil nutrients available. Bennett (1950) suggested that all soil and water conservation when properly planned and carried out on the land is wildlife conservation.

1.9.7   Time and Space

The time attribute encompasses changes in plant and animal species composition, interactions, and density (Axiom 3). Four measures of time are useful in understanding ecosystems and forest wildlife community changes: instantaneous, present, ecological, and geological.

- **Instantaneous time** is used in studying population dynamics, particularly for those species characterized by a high reproduction rate. This measure of time is concerned with rates of change of populations.
- **Present time** refers to time affecting animals in the short run, from 1 to 5 years. It is the "here-and-now" time used by managers in most management plans.
- **Ecological time** is associated with the changes in plant and animal communities and is the time most commonly used in describing plant community replacement (succession).
- **Geological time** is the tens and hundreds of thousands of years required to influence regional and global changes in topography and climate. Geological time provides a framework for understanding the past and ways of forecasting what might happen in the future.

The space attribute of natural ecosystems is not predetermined because ecosystems are open systems. Boundaries exist because humans have decided they exist, but this condition should not become a barrier to understanding and solving management problems. Space within ecosystems exists for plants and animals as they have individual boundaries that can be defined. Ecosystems can be as small as a single leaf on a tree or a fishpond or as large as a watershed or vegetation type, but the whole earth is considered the tertiary ecosystem. Locations and boundaries of ecosystems are important because they become the basis for further disaggregation for classification, inventory, monitoring, and management purposes. The concept, however, focuses on the biological, chemical, and physical processes and not specifically on space or geographical area.

1.9.8   Ecological Amplitude

Minimums, tolerances, and limiting factors are summarized in the term *ecological amplitude,* which indicates ranges of tolerances to environmental factors. Ecological amplitude was recognized by Shelford (1937) as the "law of toleration." *Ecological amplitude* has been defined in a United Nations (1997) glossary database as

> the limits of environmental conditions within which an organism can live and function.

A wide ecological amplitude suggests that a plant or animal can survive in a range of conditions, while a narrow amplitude implies specific requirements. Animals or plants with narrow ecological amplitudes are organisms most likely to be considered threatened or endangered or a candidate for one of these categories. When changes in environmental factors occur and are beyond the tolerances of any plant or animal inhabiting the local site, these plants and animals will be replaced over time; this is the process of *succession* (Axiom 3).

1.9.9 LIMITING FACTORS

The study of plant nutrition by the chemist Justus von Liebig (1840) prompted him to formulate the principle of limiting factors or the *law of the minimum*. Essentially, the law states that the nutrient in least supply is the *limiting factor* for plant growth (Axiom 1). Liebig's law of the minimum was extended to wildlife when research began to show that any one of the resource factors (food, cover, water, or space) can limit the ability of wildlife to survive. Furthermore, each species possesses a range of tolerance to its environment, meaning that there can be too much as well as too little of a specific factor such as temperature or succession (Axiom 4).

The inability to cope with any factor that exceeds the limits of such a range is often referred to as the *law of tolerance*. Any resource factor exceeding the limits of tolerance for a particular organism becomes a limiting factor for that organism. In short, all organisms require a complex composite of resources and environmental conditions and have a range of tolerance to any one of them (Odum 1971).

1.10 FOREST SUCCESSION

Succession occurs because as one plant community becomes established it slowly creates conditions that allow new species from other communities to invade, grow, and reproduce (Barbour et al. 1999) (Axiom 3). In this manner, the preceding community is gradually eliminated by a new community, a sere, where the plant species composition, abundance, and structure are different from those of the past (Figure 1.13). The last seral stage in succession is the *climax*. In addition to replacement of one

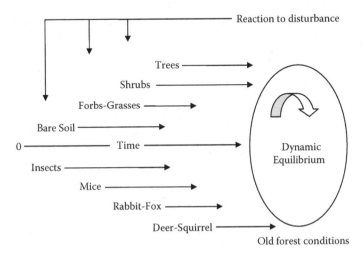

FIGURE 1.13 A general scheme of plant and animal succession and reaction to disturbance in a forest.

plant community by another, there may be instances when one plant community inhibits the establishment of the next; in this case, succession does not progress.

1.10.1   Agents of Change

Succession can be caused by either autogenic or allogenic forces. *Autogenic* forces are directed from within the ecosystem and are the agents of change caused by plants and animals. *Allogenic* succession is brought about by outside influences, for example, wind, fire, water, or human disturbance. In addition to the agents of change, the starting point of succession can be either primary or secondary. Photographic evidence of agents of change at work is contained in a publication by Gary and Currie (1977) documenting how forested areas in an abused watershed recovered over a period of 40 years.

1.10.2   Primary Succession

Primary (starting from bare soil) successional processes have a beginning, but the end may be difficult to determine. However, there comes a time in an undisturbed plant community when a balance between all the factors causing vegetation change is achieved. Under equilibrium conditions, the wastes from living matter are recycled so that the ecosystem is self-perpetuating until a major disturbance occurs; then, the process begins again (secondary succession), progressing toward climax (Knapp 1974). An example of succession that illustrates the process is from the work of Billings (1938), who studied plant species change in abandoned farms in North Carolina. He was able to document sequences of change, covering a period of 150 years, from annual plant establishment through perennial plants and grasses and transitioning to shrubs and pine to a final hardwood tree stage.

There is no disagreement by ecologists that succession occurs, but the idea of a final self-perpetuating stage, the climax, has generated much discussion in ecological literature. In general, three theories have been advanced to explain succession. The view just presented is the *monoclimax* (Barbour et al. 1999). It holds that regional climate controls the composition of the vegetation in the region; hence, the same array of plants repeatedly goes to climax via the successional process until a major climate shift occurs in the region.

A second theory (*polyclimax*) proposes that any factor (fire, soil, animals, etc.) can influence the direction of succession and array of plants in a community, not just climate (Barbour et al. 1999). As a result, the regional pattern is one of a vegetation mosaic with different seral stages leading to a different climax community. The factors influencing succession can be expressed as Conceptual Equation 1.2 (adapted from Jenny 1958) as

$$Ve = f(So + Or + To + Cl + Ti) \tag{1.2}$$

where Ve (vegetation) is a function (f) of So (soil), Or (organisms), To (topography), Cl (climate), and Ti (time).

The monoclimax and polyclimax theories assume that vegetation forms a constant and repeating identifiable and distinct unit on the ground. However, a third theory suggests that vegetation forms a continuum along environmental gradients so that individual species are replaced and combined in different ways to form communities of *climax patterns* (Whittaker 1953; Curtis 1959). This theory emphasizes that species incorporated in a climax community may change gradually or abruptly over an environmental gradient under the influence of the factors in the conceptual equation for vegetation change (Axiom 3).

Documenting and building a general model of succession has been in progress from the time of Clements (1949) until today. Walker and del Moral (2003) suggest that two schools of thought have developed in dealing with the complexity of succession: *holists* and *reductionists*. These two schools of thought are no different from the science fields of *autecology,* which is concerned with the environment of individuals, and *synecology,* which deals with the environment in a community of interacting individuals. Both approaches to studying ecosystems can contribute to the knowledge needed to manage natural resources (Lotan et al. 1981).

1.10.3 SECONDARY SUCCESSION

The concept of a stable climax based on the idea that succession is unidirectional has been challenged particularly as it relates to secondary succession (Noble and Slatyer 1977). Stable means that the climax stage persists longer than any of the other seral stages. *Secondary* succession starts when the climax stage has been disturbed either naturally or by human intervention (Axiom 7). Studies have shown that in some ecosystems (e.g., English moors and grasslands, balsam fir forests) patches of vegetation seral stages not uncommonly replace themselves in a cyclical manner (Sprugel 1975) or randomly (stochastic) in hardwood forests (Horn 1976).

Of particular interest in forest wildlife management is the gap phase replacement process identified in forest stands. In this situation, individual trees die, thereby initiating a microsuccession, which proceeds as a primary succession from bare ground. Such gaps can be small, as in the case of a single tree space, or larger, as in the case of a group of trees removed in a blowdown. The idea of gaps leads to the notion of continuing diversity even in a relatively stable climax stage. In reality, changes attributable to external factors (e.g., fire or human intervention) universally disturb climax or seral stages. Therefore, ecosystems are never stable in the true sense of the word but are resilient (Axiom 9).

1.10.4 TREE STRUCTURAL STAGES

As presented, vegetation in forested ecosystems advances through four general seral stages: bare soil, grass-forb, shrub, and tree. The tree structural stage in turn progresses through six distinct phases: seedling, sapling, young, intermediate (pole), mature, and old growth (old forest conditions) (Figure 1.14). Structure class terminology by resource agencies may change according to regional standards and vegetation type.

Succession is a powerful force that shapes vegetation structural stages, and a thorough understanding of the processes involved is critical to maintaining habitats for

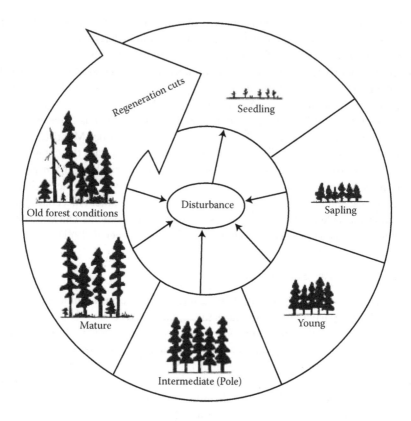

FIGURE 1.14   Tree structural stages in a forest ecosystem.

different wildlife species. Disturbances in a successional process often occur and may initiate a new stand development phase with different characteristics. Oliver and Larson (1990) suggest that these development phases follow a certain pattern as follows:

- Stand initiation
- Stem exclusion
- Understory reinitiation
- Old growth

Rate of change between stages is variable and dependent on local conditions.

The ability of humans to modify succession using the axe, plow, and cow is well documented, but we now have other and more precise means, gained by the accumulated knowledge of not only the development of plant communities but also the invention and use of tools to redirect succession over large areas in a short time (Axiom 7). Ecological knowledge together with the tools of the chain saw, bulldozer, and fire has truly made humans *ecological dominants*: organisms that by their presence can radically alter the environment of other species.

1.10.5　Old Forest Conditions

The NFMA of 1976 focused attention on the need to maintain and provide for a diversity of plant and animal communities, including old-growth forests. Forest communities aged 175–1,000 yr provide habitat for certain associated wildlife species, such as the spotted owl. Under the law, all the values found in these old forest communities must be considered, not simply their value for timber production. The old-growth forests in the Pacific Northwest became the center of attention in the 1980s due to the impact of this legislation on the increased demand for wood products from that region.

One of the major problems with old-growth issues is definition. Foresters usually consider old-growth stands as overmature or decadent stands exceeding rotation age based on the criterion of culmination of mean annual increment (MAI). Environmental groups conceive old-growth stands as possessing pristine or virgin conditions. As a result of the issues that developed in the Pacific Northwest and because no common understanding of old growth existed, the Society of American Foresters (SAF) appointed a task force to formulate an ecological definition of an average *old-growth acre* (Heinrichs 1983).

The definition applies specifically to Douglas-fir forests of the Pacific Northwest, but many of the characteristics cited are applicable to other western vegetation types. The items included relate to growth of not only a single tree but also stands and their structure and composition. The SAF task force definition states that an old-growth forest

- Consists of two or more tree species of a wide range of sizes and ages; often, a long-lived seral dominant such as Douglas-fir is associated with a shade-tolerant species such as western hemlock.
- Contains a deep, multilayered crown canopy.
- Contains more than 10 trees at least 200 yr old or 40-in dbh.
- Contains 10 snags over 20 ft tall and more than 20 tons of down logs.
- Contains at least four large snags and an equal number of logs 25-in dbh and 50 ft long.

Another definition of old growth is a broader one based on the premise that old-growth forests are relatively old and undisturbed by humans (Hunter 1989). Clearly, old growth is an ecological entity, and humans have placed a high value on preserving part or all of it for its ecological significance, wildlife habitat, or recreational and aesthetic values (Axion 8). However, several issues pertaining to old growth remain and need to be addressed by state and federal agencies for landscape-level management. Old-growth questions are as follows:

- How many acres should be conserved?
- Can silvicultural techniques duplicate existing conditions for future stands?
- At what point in time should old growth be regenerated?

The first question will take several years to answer since it involves hotly debated social and political issues. The second question cannot be addressed as easily, and it may take several generations before an answer is found. The third question will also take time because it deals with the problem of when does old growth cease to be old growth and become a decadent stand that may lose the value for which it was protected in the first place. The third question raises another question: Is old growth a renewable resource?

Kimmins (1997) made the point that the notion of a renewable resource should not be based on whether it is living or dead, but whether it can be restored to a point of reuse after a period of time. The period of time must either be within our current economic or social planning scale or be renewed at a rate that renders investment in its renewal economically attractive. Resources not meeting these criteria are classed as nonrenewable. Kimmins further observed that, with the general antipathy of the public toward cutting 1,000-yr-old redwoods in California or a 300-yr-old oak tree in France, these trees are nonrenewable on the basis of his conditions: Once cut, the social value the trees provided is gone, never to be renewed.

There are some characteristics of old-growth forests that seem to be consistent no matter what the forest type. Old trees are only one of the defining characteristics of old growth, and there is no point in time that stands obtain old-growth status, but there is a slow, long-term process of maturation that may last several hundred years. The term *old growth* does not adequately describe the total ecological meaning. Old forest conditions may represent the best descriptive term. Old forest conditions contain large trees that are past rotation age, are ecological dominants that influence site conditions, and contain dead and dying trees, and the ground layer has scattered logs and stumps in a process of advanced decay.

Old forest conditions are part of the dynamic equilibrium stage of succession (Figure 1.13). This type of forest will have small openings, with shrubs, forbs, and grasses creating an impression of being biologically diverse and with a large amount of carbon tied up in the volume of large trees and understory components. Old forest descriptive conditions will vary across the continent and forest type, and there is no one definition that fits all situations. The description of the maturation of a mature stand to old forest conditions parallels the description by Jefferies (1879) of a hollow tree destined to go before the axe. A more detailed definition of old growth can be found in *The Dictionary of Forestry* (Helms 1998).

1.10.6 Changes over Time

Coupled to plant community successional stages are the associated changes in animal species (Axiom 4). These changes occur because different animal species, like different plant species, have different life requirements and therefore different tolerances (*ecological amplitude*) to the existing environmental conditions. As plant composition changes through succession, there is in turn a change in animal species (Figures 1.15 and 1.16). The change in species is not abrupt but rather a gradual replacement process. A good record of habitat change in a forest, particularly as it affected ruffed grouse, is found in the work of Bump (1950) at the Connecticut Hill Game Management Area over a 20-yr period. Evidence is presented that as changes occurred in vegetation composition so did occupancy by the ruffed grouse.

Because of the differences in ecological amplitudes of species, many will have overlapping habitat requirements during the changes in succession stages and even

Species	Grassland	Shrubs	Pine Forest	Hardwood Forest
Grasshopper sparrow	********			
Meadowlark	**********			
Yellow-breasted chat		********		
Cardinal		********************		
Pine warbler			********	
Carolina chickadee			***************	
Crested flycatcher				*********
Acadian flycatcher				******

FIGURE 1.15 Changes in bird species as a result of succession (Axiom 4) on abandoned farmland in Georgia. (After Johnson, D. W., and E. P. Odum, *Ecology* 37:50–62, 1956.)

Species	Clearcut and Burn	Herbaceous Growth Saplings	Brush Seedlings	Mature Trees
Deer mouse	************************************			
Chipmunks		************************		
Rabbits			*********	
Red squirrel				*******
Flying squirrel				******

FIGURE 1.16 Changes in mammal species as a result of succession (Axiom 4) retarded by clearcutting and burning. (After Ream, C. H., and G. E. Gruell, Influences of Harvesting and Residue Treatments on Small Mammals and Implication for Forest Management. In *Environmental consequences for timber harvesting in Rocky Mt. coniferous forests* (1980). General Technical Report INT-90, U.S. Forest Service, Missoula, MT, pp. 455–467.)

after climax is attained. The degree of overlap is a function of ecological amplitude and niche separation. In many instances, animals that humans most desire to perpetuate, such as deer, are members of early seral rather than climax stages (Axiom 8). Thus, some wildlife species and many of the tree species (e.g., aspen) that are valuable to humans thrive best in temporary communities. Because an organism cannot survive without at least the essential parts of its requirements, wildlife managers must learn how to manipulate plant succession and so maintain the desired local communities in a more-or-less permanent state. One alternative is to allow succession to proceed normally but to manage to have a sufficient number of areas continually coming into the desired seral stages. This idea is the basic premise of silviculture and forest management.

An example of succession and how it affected one big game species can be found in the history of deer populations in the western United States (Leopold 1950). Woody shrubs, which are the staple diet of mule deer, are characteristic of early stages in forest succession. Prior to settlement of the West, deer were most abundant in the transition between forest and grassland or in areas of recent wildfires within the forest but scarce in dense timber. As forests were logged or burned, brushy plant species increased and provided the palatable shrubs necessary to maintain large deer herds.

The change in plant and animal species composition over time, on the same piece of land, results in a natural cycle of both (Axioms 3 and 4). Populations in an area may not change in total numbers of individuals but will change their location within that area. This concept is important to wildlife management because it introduces the element of *ecological time,* which in most cases is beyond the span of the *here and now* typically incorporated into management plans.

1.11 HEALTHY FOREST CONDITIONS

A major question asked by managers and researchers alike is, What is a healthy forest? In general, the traditional answer has been associated with a low number of insects and disease epidemics and a high output of timber. At one time, the lack of fire was also considered a healthy condition of a forest environment. All of these factors, however, must be considered relative to the size and intensity of the event causing the disturbance. For example, fire has a temperature range and size of burned area, insect outbreaks have an intensity of infestation as well as size of infested area, and timber harvesting can include everything from clearcutting to single tree selection. Consequently, the health of a forest is determined by the variables affecting productivity, resiliency, inhabitants, and diversity.

1.11.1 Long-Term Productivity

Forest health is related to long-term productivity and sustainability of ecosystems. In the past, the emphasis has been to produce certain amenities and commodities, particularly wood fiber, but also recreation, water, livestock forage, and wildlife habitat under a sustained-yield standard. This was done without considering how the basic ecological processes of nutrient cycling and energy flow affect the productivity and sustainability of a forest. A healthy forest is one that is resilient; that is, it has the capacity to recover from natural and human-induced changes and will in time return to a state of dynamic equilibrium.

In a healthy forest, there are different amenity and commodity potentials based on the path of ecological succession. For example, a site that is harvested for timber can, under the right circumstances, return to near-precut conditions, or a forest that becomes infested with a destructive insect can, under its own internal system or with some help from a silvicultural prescription, return to a productive state. In either case, there can be several paths to the same or different equilibrium systems.

1.11.2 Natural Events

Healthy forests will always contain endemic populations of insects and disease, a natural frequency of fire, and occasional large disturbances caused by the forces of nature similar to the loss of trees on vast acreage of forestland in South Carolina as a result of Hurricane Hugo in the September 1989 hurricane season. In addition, healthy forests should contain a continuum of age classes from seedlings to mature trees and structural features such as standing snags and large logs decaying on the ground. These features are biological legacies that provide an abundance of habitats and niches to support a large assortment of specialized plant and animal species (Franklin 1992). A more precise definition of a healthy forest will evolve when research and experience accumulate to provide baseline information for comparison of biophysical components within ecosystems to meet specific objectives within given biological constraints.

1.11.3 SUSTAINABILITY

As a result of a national concern in the late 1980s stressing the need to maintain biological diversity, the concept of a sustainable forest received attention from managers, academicians, and researchers. The emphasis placed on maintaining healthy ecosystems is one part of maintaining a sustainable forest. The term *sustainable* refers to keeping a forest in existence in some desirable condition for a given period of time. Desirable condition and period of time suggest that humans must be part of the definition and part of the process. Since human generations change over time, so will their needs and wants; therefore, the expected, predicted, or assumed desires of future generations must be protected by maintaining options for unexpected future ecosystem goods, services, and states (Bormann et al. 1993). Sustainability can be represented by a general conceptual model of the overlap of human desires to what is possible, but each area defined as an ecosystem will provide its own options for producing amenities and commodities (Figure 1.1).

Sustainability of forests is a developing concept that is receiving considerable attention because it relates to biological diversity. Measuring biological diversity and defining sustainability results in a complexity of issues of *what and how much over what period of time*. It is easy to talk in general terms about both, but how does this relate to on-the-ground management? Hagan and Whitman (2006) have proposed a way to simplify the issue by using indicators that are easy to measure and understand. The major components are economic, environmental, and social values with subcomponents of clean water, biodiversity, and other environmental values. Wildlife is considered under components of habitat diversity, game species, rare species, and other biological values. Values, subvalues, and components have indicators that relate to condition, pressure, and policy response.

1.11.4 CLIMATE CHANGE

Climate has a strong influence on establishment of the vegetation of the world and associated wildlife and has been in constant change over geological time. Weather and climate are physical environmental factors and at times are used as synonyms, but there is a difference. *Weather* refers to the day-to-day combinations of temperature, moisture, barometric pressure, and wind. *Climate* is a description of long-term weather patterns in specific regions or areas. Variations in life forms (forests, grassland, etc.) of vegetation structure are controlled primarily by temperature and precipitation as they are influenced by latitude and topography (Conceptual Equation 1.2). The tilt of the earth and its rotation around the sun defines the weather changes by day and season and long-term climate changes over ecological time. Whittaker (1970) found a relationship between mean annual temperature and mean annual precipitation by plant type (biome). The change in biome life forms results from the amount of solar radiation received by plants at the surface of the earth, and this is where the concern about climate change is currently focused.

The earth has always had climate change over geological time, but is there a change or can there be a change in ecological time of decades or centuries as a result of human use of nonrenewable resources and manufactured chemicals? The concern is whether there is an increase in gasses, primarily CO_2, that produce a "greenhouse" effect followed by a rise in temperature. What are the potential effects of a rise in mean annual temperature if it occurs over ecological time? Some of the consequences could be as follows:

- Changes in the biology of animals such as migration patterns, breeding seasons, geographical distribution, and gestation or hatching time
- A reduction in the survival of young that cannot adjust to heat
- Changes in plant species composition, structure, and abundance, which are habitat components
- Changes in size and distribution of forest types

The four items listed are not all inclusive, but there are many indirect effects that could be detrimental to a particular species. The question is, What are examples that indicate a climate change has taken place or is anticipated in the short term? The most recent figures available indicate the following (Climate Change Science Program [CCSP] 2008; Intergovernmental Panel on Climate Change [IPCC] 2007):

- There has been a temperature increase of 0.74°C in global average surface temperature over the last 100 yr.
- Eleven of the last 12 yr are among the 12 warmest years since the instrumental record of global surface temperature was started in 1850.
- In the contiguous United States, the increase in total annual precipitation has been 6.1% over the last century.
- The patterns of precipitation are variable, but increases of more than 10% have occurred in the northeast and South.
- Since 1960 in the western United States, more precipitation has been falling as rain than snow (Knowles et al. 2006).

The examples listed are a few of the known facts about temperature and changes in precipitation. Other documented examples are warming of the oceans, a rising sea level, changes in storm intensity, changes in ocean pH, warming in the Arctic, changes in extreme events, and changes in hydrology (CCSP 2008). How does a change in temperature translate to an actual cause-effect case at the management level? One example is the polar bear, which uses sea ice as a platform for hunting seals, mating activities, and seasonal movements (Sterling and Derocher 2007). Since about 1970 there has been a trend to an earlier breakup of sea ice, and this has affected the polar bear in Western Hudson Bay. Polar bears have been forced to abandon hunting on sea ice at the most important time of year—when seal pups are abundant and not too aware of predators. The result is that bears have to move inland and fast until the ice forms again. A shortened time for feeding on seals results in a weight loss for bears, which affects their condition and biological functions such as mating and the rearing of cubs.

Global warming and climate change are now a major political issue at both the national and international levels and will continue to be an issue for years to come. The problem for managers is not only how to adjust to slow climate change but also how to adapt practices to the consequences of extreme events such as fires, earthquakes, and storms that seem to be more common. Adaptive forest management is most certainly one approach to the problem.

1.12 THE MENOMINEE PARADIGM

Historically, the Menominee Indians are the oldest in Wisconsin, and the tribe once owned 9.5 million acres of land in the central and mideastern part of the state. The reservation was terminated in 1961 but was restored in 1973 with ownership of 230,000 ac. A forest management plan for the reservation was approved by the tribal legislature and Bureau of Indian Affairs (BIA) in 1988. The forest management plan is a true forest plan and does not consider the effects of management practices on other natural resources, such as wildlife, fish, recreation, aesthetics, or cultural and religious values.

A system of sustained-yield forest management using selection harvesting was initiated by the tribal leaders after the reservation was established in 1854. The vision was one of a forest to be harvested at a rate that would achieve a balance between growth, mortality, and the annual production of timber. The Menominees believe that resource development must be part of a larger social and economic plan rather than an end in itself. At present, the tribe does not have management plans for nonconsumptive uses of their forest resource (e.g., nongame wildlife). They believe that good forest management practices based on sustained yield determined by a continuous forest inventory will produce diverse conditions and good wildlife habitat, water quality, and aesthetic values (Pecore 1992).

Available figures support the Menominee's claim of sustained yield. In 1854, there were approximately 1.5 billion board feet of saw timber growing stock. An inventory in 1979 indicated that the saw timber stock remained at 1.5 billion board feet. Over a 128-yr period, more than 2 billion board feet has been harvested. Annually, the tribe harvests approximately 24 million board feet of saw timber and 55,000 cords of pulpwood. The success of the Menominee tribe in maintaining a sustained yield is based on a philosophy that has not changed over time: a well-regulated forest using area and volume control with even- and uneven-aged management, a large number (884) of continuous forest inventory (CFI) plots contained in 109 compartments, and limits on the allowable cut not to exceed 7% annually, 3% for a 5-yr period, and 2% over a 15-yr cutting cycle.

Forestry has been the major force setting the direction for Menominee resource management. The current members have inherited a forest that has not been abused, so there have not been any major wildlife problems noticeable to the tribe. As the human population increases, there could be more demand for certain featured species, such as deer, which probably will remain at a constant density level; therefore, the same number of deer will be hunted by a larger human population. For years, since the 1930s and 1940s, foresters have been saying that good forestry is good wildlife management, but many on-the-ground examples have proven otherwise. If a truly regulated forest could be found, then the paradigm of *good forest management is good multispecies wildlife management* would probably be true. The Menominee reservation may be the closest to this paradigm.

1.13   THE PRINCIPLE OF COMPLEMENTARITY

Practicing natural resource managers who make day-to-day decisions must be totally frustrated by the concepts and terminology that they have to understand for planning purposes and field application. However, answers to these frustrations can often be found outside our own profession. In the case of terminology, such as new perspectives, biodiversity, ecological restoration, matrix management, conservation biology, and habitat relationships, there may be an answer in physics.

Neils Bohr, an atomic physicist, concluded that information obtained under one set of experimental conditions might not be the same as or even consistent with information obtained under a different set of procedures (Weaver 1975). In such cases, the information obtained from the first experiment must be viewed as complementary to the information obtained in the second. Even if the two sets of data appear contradictory, both must be accepted as equally valid if no errors exist in the data. For example, under some experimental conditions, electrons behave as particles, but under others they behave as waves as in light. The question becomes, Can electrons be both particles and waves? The *principle of complementarity* asserts that they can. By analogy, Bohr's principle can mean that different approaches to solving a common problem, such as in natural resource management, are equally valid if the end results are the same.

In the conservation of natural resources, we deal with animals and their populations, plants as habitat, and how both can be sustained over time. The goal is to manage without any animal becoming extinct as a result of human activity. Biodiversity, conservation biology, habitat relationships, and so on are only ways to address a common problem, and all are valid processes to be considered in the management, planning, and decision-making process because they are different approaches to reach the same conclusion. The principle of complementarity allows managers to select their management system or practices to conform to local conditions, including public opinion, as long as the end product meets the objectives of their management plan.

KNOWLEDGE ENHANCEMENT READING

Barbour, M. G., and D. W. Billings. 2000. *North American terrestrial vegetation.* Cambridge University Press, UK.

Barbour, M. G., J. H. Burk, W. D. Pitts, et al. 1999. *Terrestrial plant ecology.* Addison-Wesley, Longman, New York.

Boyce, M. S., and A. Haney. 1997. *Ecosystem management: Application for sustainable forest and wildlife resources.* Yale University Press, New Haven, CT.

Hocker, H. W., Jr. 1979. *Introduction to forest biology.* John Wiley and Sons, New York.

Hunter, M. L. Jr. 1990. *Wildlife, forests and forestry: Principles of managing forests for biological diversity.* Prentice-Hall, Englewood, NJ.

Kimmins, J. P. 1996. *Forest ecology.* Prentice-Hall, Upper Saddle River, NJ.

Krausman, P. R. 2002. *Introduction to wildlife management.* Prentice-Hall, Upper Saddle River, NJ.

Lindenmayer, D. B., and J. F. Franklin. 2002. *Conserving forest biodiversity: A comprehensive multiscaled approach.* Island Press, Washington, DC.

Perry, D. A., R. Oren, and S. C. Hart. 2008. *Forest ecosystems.* Johns Hopkins University Press, Baltimore, MD.

Smith, T. M., and R. L. Smith. 2008. *Elements of ecology.* Benjamin Cummings, New York.

Walker, L. R., and R. del Moral. 2003. *Primary succession and ecosystem rehabilitation.* Cambridge University Press, Cambridge, UK.

Young, R. A., and R. Giese (Eds.). 2003. *Introduction to forest ecosystem science and management.* John Wiley and Sons, Hoboken, NJ.

REFERENCES

Allen, D. L. 1974. *Our wildlife legacy.* Funk and Wagnalls, New York.

Andrewartha, H. G., and L. C. Birch. 1984. *The ecological web: More on the distribution and abundance of animals.* University of Chicago Press, Chicago.

Atwater, S., and J. Schnell (Eds.). 1989. *Ruffed grouse.* Stackpole Books, Harrisburg, PA.

Bailey, J. A. 1984. *Principles of wildlife management.* John Wiley and Sons, New York.

Barbour, M. G., and D. W. Billings. 2000. *North American terrestrial vegetation.* Cambridge University Press, Cambridge, UK.

Barbour, M. G., J. H. Burk, W. D. Pitts, et al. 1999. *Terrestrial plant ecology.* Addison-Wesley, Longman, New York.

Bennett, G. W. 1986. *Management of lakes and ponds.* Robert E. Krieger, Malabar, FL.

Bennett, H. 1950. Soil conservation and wildlife. Paper presented at 13th annual convention, June 17. Michigan United Conservation Clubs, Ludington, MI.

Billings, W. D. 1938. The structure and development of old field shortleaf pine stands and certain associated physical factors of soil. *Ecol. Monogr.* 8:437–499.

Bissonette, J. A. (Ed.). 1997. *Wildlife and landscape ecology: Effects of pattern and scale.* Springer-Verlag, New York.

Block, W. M. 1993. *The habitat concept in ornithology, theory and applications: Current ornithology.* Vol. 2. Plenum Press, New York.

Borman, B. T., P. G. Cunnigham, M. H. Brooks, et al. 1993. Adaptive ecosystem management in the Pacific northwest. General Technical Report PNW–GTR–341. U.S. Forest Service, Portland, OR.

Boyce, S. G., and N. D. Cost. 1978. *Forest diversity: New concepts and applications.* Research Paper SE-194. U.S. Forest Service, Asheville, NC.

Brown, D. E. (Ed.). 1998. *A classification of North American biotic communities.* University of Utah Press, Salt Lake City.

Brown, J. K., E. D. Reinhardt, and K. A. Kramer. 2003. *Course woody debris: Managing benefits and fire hazards in the recovering forest.* General Technical Report 129. U.S. Forest Service, Ogden, UT.

Bump, G. 1950. *Wildlife habitat changes in the Connecticut Hill game management area: Their effect on certain species of wildlife over a twenty-year period.* Cornell University Agricultural Experimental Station, Ithaca, NY.

Burns, R. M., and B. H. Honkala. 1990. *Silvics of trees of North America.* Vol. 1, *Conifers,* Vol. 2, *Hardwoods.* Agriculture Handbook 654. U.S. Forest Service, Washington, DC.

Cheatum, E. L., and C. W. Severinghaus. 1950. Variations in fertility of white-tailed deer related to range conditions. *Trans. N. Am. Wildl. Conf.* 15:170–189.

Clary, W. P. 1978. *Producer-consumer biomass in Arizona ponderosa pine.* General Technical Report RM-56. U.S. Forest Service, Fort Collins, CO.

Clements, F. E. 1949. *Dynamics of vegetation.* H. W. Wilson, New York.

Climate Change Science Program (CCSP). 2008. *Preliminary review of adaption options for climate-sensitive ecosystems and resources.* Julius, S. H. and J. M. West, Eds. A report by the U.S. Climate Change Science Program and the Subcommittee on Global Change Research, U.S. Environmental Protection Agency, Washington, DC.

Cody, M. L., and J. M. Diamond (Eds.). 1975. *Ecology and evolution of communities.* Harvard University Press, Cambridge, MA.

Cohen, W. B., M. E. Harmon, D. O. Wallin, et al. 1996. Two decades of carbon flux from forests of the Pacific Northwest. *BioScience* 46:836–843.

Colinvaux, P. 1986. *Ecology.* John Wiley and Sons, New York.

Crawford, W. T. 1950. Some specific relationships between soils and wildlife. *J. Wildl. Manage.* 30:396–398.

Curtis, J. P. 1959. *The vegetation of Wisconsin.* University of Wisconsin Press, Madison.

Dasmann, R. F. 1981. *Wildlife biology.* John Wiley and Sons, New York.

Dayton, W. A. 1931. *Important western browse plants.* Miscellaneous Publication 101, U.S. Department of Agriculture. Government Printing Office, Washington, DC.

Dice, L. R. 1943. *The biotic provinces of North America.* University of Michigan Press, Ann Arbor.

Emmel, T. C. 1973. *An introduction to ecology and population biology.* W. W. Norton and Company, New York.

Everhart, W. H., and W. D. Youngs. 1981. *Principles of fishery science.* Cornell University Press, Ithaca, NY.

Eyre, F. H. (Ed.). 1980. *Forest cover types of the United States and Canada.* Society of American Foresters, Washington, DC.

Forbs, R. D. (Ed.). 1961. *Forestry Handbook.* Forest Wildlife Management Working Group, Section 9, pp. 38–39.

Ford-Robertson, F. C. 1983. *Terminology of forest science technology practice and products.* FAO/IUFRO Committee on Forestry Bibliography and Terminology. Society of American Foresters, Washington, DC.

Franklin, J. F. 1992. Scientific basis for new perspectives in forests and streams. In R. J. Naiman (Ed.), *Watershed management: Balancing sustainability and environmental change,* pp. 25–72. Springer-Verlag, New York.

Franklin, J. F., and R. T. T. Forman. 1987. Creating landscape patterns by forest cutting: Ecological consequences and principles. *Landscape Ecol.* 1:5–18.

French, C. E., L. C. McEwen, N. D. MacGruder, et al. 1956. Nutrient requirements for growth and antler development in the white-tailed deer. *J. Wildl. Manage.* 20:221–232.

Garrison, G. A., A. Bjugstad, D. A. Duncan, et al. 1977. *Vegetation and environmental features of forest and range ecosystems.* Agriculture Handbook 475. U.S. Forest Service, Washington, DC.

Gary, H. L., and P. O. Currie. 1977. *The Front Range pine type: A 40-year photographic record of plant recovery on abused watershed.* General Technical Report RM-46. U.S. Forest Service, Fort Collins, CO.

Gaston, K. J., and J. I. Spicer. 2004. *Biodiversity: An introduction.* Blackwell Scientific, Oxford, UK.

Gates, D. M. 1965. Radiant energy, its receipt and disposal. *Meterol. Monogr.* 6:1–26.

Goodman, D. 1975. The theory of diversity-stability relationships in ecology. *Q. Rev. Biol.* 50:237–266.

Hagan, J. M., and A. A. Whitman. 2006. Biodiversity indicators for sustainable forestry; simplifying complexity. *J. Forest.* 104:203–210.

Hall, L. A., P. R. Krausman, and M. L. Morrison. 1997. The habitat concept and a plea for standard terminology. *Wildl. Soc. Bull.* 25(1):173–182.

Halls, L. K. 1984. *White-tailed deer: Ecology and management.* A Wildlife Management Institute Book. Stackpole Books, Harrisburg, PA.

Harmon, M. E., J. F. Franklin, F. J. Swanson, et al. 1986. Ecology of coarse woody debris in temperate ecosystems. *Adv. Ecol. Res.* 15:133–302.

Harris, L. D. 1984. *The fragmented forest: Island biogeography theory and the preservation of biological diversity.* University of Chicago Press.

Harvey, A. E., M. F. Jurgensen, M. J. Larsen, et al. 1987. *Decaying organic materials and soil quality in the Inland Northwest: A management opportunity.* General Technical Report INT-225. U.S. Forest Service, Odgen, UT.

Heinrichs, J. 1983. Old growth comes of age. *J. Forest.* 81:776–779.

Helms, J. A. (Ed.). 1998. *The dictionary of forestry.* Society of American Foresters, Bethesda, MD.

Hickman, C. P. 1966. *Integrated principles of zoology.* C.V. Mosby, St. Louis, MO.

Hill, E. P. 1972. *The cottontail in Alabama.* Bulletin No. 440, Alabama Agriculture Experimental Station. Auburn University, Alabama.

Holling, C. S. 1973. Resilience and stability of ecological systems. *Annu. Rev. Ecol. Syst.* 4:1–23.

Horn, H. S. 1976. Succession. In R. M. May (Ed.), *Theoretical ecology, principles and applications,* pp. 187–204. Blackwell Scientific, Oxford University Press, Oxford, UK.

Houghton, R. A. 1995. Land-use changes and the carbon cycle. *Global Change Biol.* 1:275–287.

Hunter, M. L., Jr. 1989. What constitutes an old-growth stand? *J. Forest.* 87:33–35.

Hutchinson, G. E. 1959. Homage to Santa Rosalia, or why are there so many kinds of animals? *Am. Nat.* 93:145–159.

Intergovernmental Panel on Climate Change (IPCC). 2007. *Climate change 2007: The physical science basis.* Contribution of Working Group I to the Fourth Assessment Report of the Intergovernmental Panel on Climate Change. Cambridge University Press, Cambridge, UK.

Jefferies, R. 1879. *The gamekeeper at home: Sketches of natural history and ranch life* (Third edition). Roberts Brothers, Boston, MA.

Jenkins, D. H., and I. H. Bartlett. 1959. *Michigan whitetails.* Michigan Department of Conservation, Lansing.

Jenny, H. 1958. Role of the plant factor in the pedogenic functions. *Ecology* 39:6–16.

Johnson, D. W., and E. P. Odum. 1956. Breeding bird populations in relation to plant succession on the Piedmont of Georgia. *Ecology* 37:50–62.

Johnson, R. 1970. *Tree removal along southwestern rivers and effects on associated organisms.* American Philosophical Society Yearbook, Philadelphia, PA, pp. 431–432.

Johnson, R. R., and D. A. Jones (Tech. Coords.). 1977. *Importance, preservation and management of riparian habitat.* General Technical Report RM-42. U.S. Forest Service, Fort Collins, CO.

Johnson, R. R., and C. D. Ziebell (Tech. Coords.). 1985. *Riparian ecosystems and their management: Reconciling conflicting uses.* General Technical Report RM-120. U.S. Forest Service, Fort Collins, CO.

Kimmins, J. P. 1997. *Forest ecology: A foundation for sustainable management.* Macmillan, New York.

Klopfer, P. H. 1969. *Habitats and territories: A study of the use of space by animals.* Basic Books, New York.

Knapp, R. 1974. *Cyclic successions and ecosystem approaches in vegetation dynamics.* Handbook of Vegetation Science. W. Junk, Hague, Netherlands.

Knowles, N., D. Dettinger, and D. R. Cayan. 2006. Trends in snow fall versus rainfall in Western United States. *J. Climate* 19(18):4545–4559.

Kuchler, A. W. 1966. *Potential natural vegetation* [map]. University of Kansas. U.S. Forest Service, Washington, DC.

Lee, R. M. 1989. *The desert bighorn sheep in Arizona.* Arizona Game and Fish Department, Phoenix.

Leopold, A. 1933. *Game management.* Charles Scribner's Sons, New York.

Leopold, A. S. 1950. Deer in relation to plant succession. *Trans. N. Am. Wildl. Conf.* 15:571–580.

Lewin, R. 1986. In ecology, change brings stability. Research News. *Science* 234:1071–1073.

Lindenmayer, D. B., and J. F. Franklin. 2002. *Conserving forest biodiversity: A comprehensive multiscaled approach.* Island Press, Washington, DC.

Lotan, J. E., M. Alexander, S. F. Arno, et al. 1981. *Effects of fire on flora.* General Technical Report WO-16. U.S. Forest Service, Washington, DC.

MacArthur, R. H. 1965. Patterns of species diversity. *Biol. Rev.* 40:510–533.

Martin, A. C., H. S. Zim, and A. L. Nelson. 1951. *American wildlife and plants, a guide to wildlife food habits.* Dover, New York.

May, R. M. 1973. *Stability and complexity in model ecosystems.* Princeton University Press, Princeton, NJ.

Meehan, W. R., F. J. Swanson, and J. R. Sedell. 1977. Influences of riparian vegetation on aquatic ecosystems with particular reference to salmonid fishes and their food supply. In R. Johnson and D. A. Jones (Tech. Coords). 1977. *Importance, preservation and management of riparian habitat.* General Technical Report RM–42. U.S. Forest Service, Fort Collins, CO.

Merriam, C. H. 1898. *Life zones and crop zones of the United States.* U.S. Biological Survey Bulletin 10:1–79. Government Printing Office, Washington, DC.

Millspaugh, J. J., and F. R. Thompson, III. 2009. *Models for planning wildlife conservation in large landscapes.* Academic Press, San Diego, CA.

Morrison, M. L. 2002. *Wildlife restoration, techniques for habitat analysis and animal monitoring.* Society for Ecological Restoration. Island Press, Washington, DC.

Noble, I. R., and R. O. Slatyer. 1977. Post-fire succession of plants in Mediterranean ecosystems. In H. A. Money and C. E. Conrad (Eds.), *Environmental consequences of fire and fuel management in Mediterranean ecosystems,* pp. 27–63. General Technical Report WO-3. U.S. Forest Service, Washington, DC.

Odum, E. P. 1971. *Fundamentals of ecology.* W. B. Saunders, Philadelphia.

Office of Technical Assessment. 1987. *Technologies to maintain biological diversity.* Congress of the United States. U.S. Government Printing Office, Washington, DC.

Oliver, C. D., and B. C. Larson. 1990. *Forest stand dynamics: Biological resource management series.* McGraw-Hill, New York.

O'Neill, R. V., D. L. DeAngelis, J. B. Waide, et al. 1986. *A hierarchical concept of ecosystems.* Monograph in Population Biology-23. Princeton University Press, Princeton, NJ.

Patton, D. R. 1992. *Wildlife habitat relationships in forested ecosystems.* Timber Press, Portland, OR.

Pecore, M. 1992. Menominee sustained-yield management: A successful land ethic in practice. *J. Forest.* July:12–16.

Pimlot, D. H. 1969. The value of diversity. *Trans. N. Am. Wildl. Nat. Resource Conf.* 34:265–280.

Primack, R. B. 2006. *Essentials of conservation biology.* Sinauer Associates, Sunderland, MA.

Ream, C. H., and G. E. Gruell. 1980. Influences of harvesting and residue treatments on small mammals and implication for forest management. In *Environmental consequences for timber harvesting in Rocky Mt. coniferous forests,* pp. 455–467. General Technical Report INT-90. U.S. Forest Service, Missoula, MT.

Ripley, B. D. 1981. *Spatial statistics.* John Wiley and Sons, New York.

Ross, M. R. 1997. *Fisheries conservation and management.* Prentice-Hall, Upper Saddle River, NJ.

Schwartz, B. 2004. *The paradox of choice—why more is less: How the culture of abundance robs us of satisfaction.* Harper Collins, New York.

Shantz, H. L., and R. Zon. 1924. Forest vegetation of the United States. In *Natural vegetation: Atlas of American agriculture.* Agriculture Handbook 271. U.S. Department of Agriculture, Washington, DC.

Shelford, V. E. 1937. Animal communities in temperate America. The Geographic Society of Chicago, Bulletin No. 5:1–368. The University of Chicago Press, Chicago, IL.

Shelford, V. E. 1963. *The ecology of North America.* University of Illinois Press, Urbana.

Siderits, K., and R. E. Radtke. 1977. Enhancing forest wildlife habitat through diversity. *Trans. N. Am. Wildl. Nat. Resource Conf.* 42:425–433.

Skalski, J. R., and D. S. Robson. 1992. *Techniques for wildlife investigations: Design and analysis of capture data.* Academic Press, Harcourt Brace Jovanovich, New York.

Skinner, T. H., and J. O. Klemmedson. 1978. *Abert squirrel influences nutrient transfer through litterfall in a ponderosa pine forest.* Research Note RM-353. U.S. Forest Service, Fort Collins, CO.

Smith, C. M., E. D. Michael, and H. V. Wiant Jr. 1975. Size of West Virginia deer as related to soil fertility. *W. Va. Agric. Forest.* 6:12–13.

Smith, R. L. 1990. *Ecology and field biology.* Harper and Row, New York.

Smith, R. L., and T. M. Smith. 2009. *Elements of ecology.* Benjamin Cummings, New York.

Soule, M. E. (Ed.). 1986. *Conservation biology: The science of scarcity and diversity.* Sinauer Associates, Sunderland, MA.

Sprugel, D. G. 1975. Dynamic structure of wave-regenerated Abies balsamia forests in the northeastern United States. *J. Ecol.* 64:889–911.

States, J. S. 1979. *Squirrel-trees-truffles: A model of interactive dependency.* Progress Report, Ecological Studies of Hypogeous Fungi, Project Number 4. Department of Biological Science, Northern Arizona University, Flagstaff.

Sterling, I., and A. E. Derocher. 2007. Melting under pressure: The real scoop on climate warming and polar bears. *Wildl. Prof.* 43:23–27.

Tansley, A. G. 1935. The use and abuse of vegetational concepts and terms. *Ecology* 16:284–307.

Tarbell, H. S. 1899. *The complete geography.* American Book Company, New York.

Thomas, J. W. (Ed.). 1979. *Wildlife habitats in managed forests: The Blue Mountains of Oregon and Washington.* Agriculture Handbook 553. U.S. Forest Service, Washington, DC.

Thomas, J. W., and D. E. Toweill. 1982. *Elk of North America ecology and management.* A Wildlife Management Institute Book. Stackpole Books, Harrisburg, PA.

Thomas, J. W., C. Maser, and J. E. Rodick. 1979. Edges. In J. W. Thomas (Ed.), *Wildlife habitats in managed forests; The Blue Mountains of Oregon and Washington,* pp. 47–59. Agriculture Handbook 553, U.S. Forest Service, Washington, DC.

Trappe, J. M., and C. Maser. 1978. Ectomycorrhizal fungi: Interaction of mushrooms and truffles with beasts and trees. In T. Walters (Ed.), *Mushrooms and man: An interdisciplinary approach to mycology*, pp. 163–179. Linn-Benton Community College, Albany, OR.

Turner, M. G., R. H. Gardner, and R. O. O'Neill. 2001. *Landscape ecology in theory and practice: Pattern and process*. Springer-Verlag, New York.

United Nations. 1973. *International classification and mapping of vegetation*. Ecology and Conservation Series 6. United Nations Educational, Scientific, and Cultural Organization (UNESCO), Paris.

United Nations. 1997. *Studies in methods: Glossary of environmental statistics*. Series F, No. 67. United Nations Statistical Division: New York.

U.S. Congress, 1974. Forest and Rangeland Renewable Resources Planning Act, Public Law 93–378, U.S. Government Printing Office, Washington, DC.

U.S. Congress, 1976. National Forest Management Act, Public Law 94–988, U.S. Government Printing Office, Washington, DC.

U.S. Department of Agriculture (USDA). 1948. *Forest regions of the United States* [map]. U.S. Forest Service, Washington, DC.

U.S. Department of Agriculture (USDA). 1967. *Major forest types* [map]. National Atlas. U.S. Forest Service, Washington, DC.

U.S. Department of Agriculture (USDA). 1974. *Forest and range ecosystems of the United States* [map]. U.S. Forest Service, Washington, DC.

U.S. Department of Agriculture (USDA). 1990. *Conserving our heritage, Americas biodiversity*. U.S. Forest Service, Washington, DC.

U.S. Department of Agriculture (USDA). 2000. *Forest cover types* [map]. The National Atlas of the United States of America. Compiled by U.S. Forest Service, Washington, DC, and U. S. Geological Survey, Denver, CO.

Vannote, R. L., G. W. Minshall, K. W. Cummins, et al. 1980. The river continuum concept. *Can. J. Fish. Aquat. Sci.* 37(1):130–137.

von Liebig, J. 1840. *Chemistry in its application to agriculture and physiology*. Taylor and Walton, London.

Walker, L. C. 1999. *The North American forests: Geography, ecology, and silviculture*. CRC Press, Taylor and Francis Group, New York.

Walker, R. W., and R. del Moral. 2003. *Primary succession and ecosystem rehabilitation*. Cambridge University Press, Cambridge, UK.

Weaver, W. 1975. The religion of a scientist. In L. Rosten (Ed.), *Religions of America*, pp. 296–305. Simon and Schuster, New York.

Welch, P. S. 1955. *Limnological methods*. McGraw-Hill, New York.

Whittaker, R. H. 1953. A consideration of climax theory: The climax as a population and pattern. *Ecol. Monogr.* 23:41–78.

Whittaker, R. H. 1970. *Communities and ecosystems*. Macmillan, New York.

Yeager, L. E. 1961. Classification of North American mammals and birds according to forest habitat preference. *J. Forest.* 9:671–674.

Young, R. A., and R. L. Giese (Eds.). 2003. *Introduction to forest ecosystem science and management*. John Wiley and Sons, Hoboken, NJ.

Zhu, Z., and Evans, D. L. 1994. *U.S. forest types and predicted percent forest cover from AVHRR data*. U.S. Forest Service, U.S. Geological Survey, Washington, DC.

2 Survive and Reproduce

2.1 INTRODUCTION

In Chapter 1, a case was made that six environmental factors in the centrum of an animal affect its chance to survive and reproduce (Conceptual Equation 1.1). In the paragraphs that follow, these six factors are reviewed to provide a better understanding of how each contributes to the environment and habitat of animals that live in a forest ecosystem. A discussion of these factors can provide wildlife biologists with insight about the complexity of the resources they manage. By keeping these factors in mind in the planning and assessment process, it may be possible to identify and correct potential problems before they occur.

2.2 HAZARDS

A *hazard* is anything (excluding disease) capable of causing injury or death to an animal. Wildlife must cope with both natural and man-made hazards. Natural hazards include such things as water, snow, wind, and fire. Man-made hazards include roads, motorized vehicles, fences, and intentional or accidentally caused fire. In a broader context, hazards could also include humankind's hunting activity and habitat destruction. Probably the most common and greatest threat to wildlife from natural and man-made hazards is from fire, roads, and fences.

2.2.1 FIRE

Fire has both direct and indirect effects. The direct effect is death by burning as well as suffocation, particularly in the case of small mammals (Lawrence 1966). The air temperature at which animals are killed is approximately 145°F (Howard et al. 1959). Direct mortality has been documented in animals as varied as birds and voles (Bock and Lynch 1970; Hakala et al. 1971), deer, wood rats, and rabbits (Chew et al. 1958; Buech et al. 1977). In most cases, even in large fires, death by burning is rare (Vogl 1977). In general, the direct loss of vertebrates to the effects of fire is negligible, except in the case of large, hot fires. The reason for this could be a behavioral trait, such as for reptiles, which have heat-sensing pits they use to locate prey (Howard et al. 1959) or for amphibians to seek wet mud until the flames have decreased (Komarek 1969).

The impact of fire on an animal is a function of the size and mobility of the animal and the extent, intensity, and speed of the fire. Adult deer and elk commonly avoid injury because they can move away easily from advancing fire, but very young animals may not be so fortunate. Chipmunks, shrews, wood rats, and mice are generally reluctant to leave burning woodpiles and grass stands; therefore, they are probably killed directly (Bendell 1974). However, those species living underground can survive intense fire due to the great decrease in temperature below the surface of the ground (Howard et al. 1959).

The indirect effects of fire on wildlife, although difficult to quantify, are substantially greater due to the loss of habitat, which leads to starvation, predation, or decimation by weather. These indirect effects usually persist for 1–3 yr (Cook 1959; Lawrence 1966). Starvation following fires that burn large areas results from the loss of vegetation used for food, particularly in the case of species of limited mobility (home range). Even when starvation does not result in death, the ability of an animal to reproduce is frequently severely impacted because, when food is limited, nutrients are used for body maintenance rather than reproduction.

Predation on small mammals following fire is generally increased since a survivor has to search for food at greater distances from cover. Fire may reduce vegetation to an extent that animals are frequently exposed in open areas, thereby increasing their vulnerability to predators (Edwards 1954). It is not uncommon to see hawks and owls in recently burned areas, presumably because small mammals are easier to hunt in the open spaces where there is little or no vegetation. Decimation of animals from weather after fire results from a loss of cover that otherwise provided protection from wind, water, and heat. Although general cover may still exist in the burned area, the special types of vegetative cover needed for nesting, feeding, resting, and loafing may be too thin to protect both adults and young from the elements (Baker 1974).

Little information is available on the effects of fire on amphibians, and the information with respect to reptiles is just as limited. Amphibians have been observed to remain in mud and water until a fire has passed (Komarek 1969). Reptiles may be able to escape due to their ability to detect fire with their heat-sensing pits (used in locating prey). Birds are most vulnerable during nesting and fledging periods. Fire devastates ground-nesting birds not only because it destroys nests, but also because it removes protective cover and eliminates insects used for food (Daubenmire 1968).

The impacts of fire on fish are indirectly the result of physical and chemical changes caused by vegetation removal in the surrounding watershed. Removal of stream bank vegetation commonly increases water temperatures, thereby converting a cold-water stream into one of warm water, leading to corresponding changes in species composition (Lyon et al. 1978; Swanston 1980). A change in water temperature may also affect fish distribution patterns (Koya 1977). Fire generally results in an increase in erosion, leading to increased sedimentation, killing fry or destroying eggs (Swanston 1980).

Because fire can have both detrimental and beneficial effects on wildlife species, there is often a need to assess the benefits and losses for management and planning purposes. A response classification system that was developed by Walter (1977) has four categories to associate wildlife species with fire:

- **Fire-intolerant** species that decrease in abundance after fire
- **Fire-impervious** wildlife species that are unaffected by burning

- **Fire-adapted** species that are associated with habitats characterized by recurring fires of various intensities
- **Fire-dependent** species that are associated with fire-dependent and fire-adapted plant communities

These four categories were used to place a number of wildlife species into four categories based on current literature at the time (Kramp et al. 1983). A few examples show how selected species respond to fire in the Southwest:

- **Intolerant:** Brown towhee, red squirrel, western flycatcher
- **Impervious:** Cedar waxwing, eastern kingbird, eastern cottontail
- **Adapted:** House wren, Gambel's quail, bobcat
- **Dependent:** Blue grouse, mourning dove, elk

As with any species, the ecological amplitude may result in an animal being listed in more than one category, but this does not lessen the usefulness of the system developed by Walter.

2.2.2 ROADS

Each year, the number of miles of paved highways and improved dirt roads in forested areas increases. About 18% of the nation's 3.8 million miles of roads are located within cities and towns, while the remaining 82% are in rural areas where the chance for encountering wildlife is increased (Council on Environmental Quality 1981). The killing of animals along a highway is a direct hazard. Mortality and collisions with vehicles is well documented, and accidents kill many species of wildlife (Trombulak and Frissell 2000). Statistics show the loss in terms of wildlife and humans from reported vehicle collisions with animals ("Roadkill Statistics" 2005):

- 3.8 million miles of roads in the United States
- 6.3 million auto accidents annually in the United States
- 253,000 animal-vehicle accidents annually
- 90% of animal-vehicle collisions involve deer
- 1 million vertebrates are run over each day in the United States
- 200 humans die each year from vehicle-wildlife collisions

The figures presented for animal-vehicle collisions are probably low because many accidents go unreported.

There are indications that some species have a tolerance level to roads so that flight does not lead to displacement (Lyon 1984). Data from a study in Wyoming on the effects of logging and recreational road traffic on elk indicate little effect at a distance of about 0.6 mi from timber harvest operations, and after activity stops, animals move back into the harvested area (Ward 1976). The same study concluded that elk were little affected by logging and recreation roads during the day within 0.25 mi once the elk became used to the traffic. Elk seem to avoid areas of heavy vehicle traffic, but the level of avoidance depends on the extent of cover (Perry and Overly 1976; Lyon 1984). Indirect effects arise from the noise, disturbance, and fragmentation that may prevent animals such as deer and elk from using an area.

In the Southwest, there is a strong association between riparian meadow habitats and elk in east central Arizona. This in turn is reflected in a substantially higher degree of elk crossing roads and elk-vehicle collisions adjacent to riparian habitats (Dodd et al. 2006). Given the increasing human population and volume of automobile traffic in the area, elk crossing roads at specific sites are likely to become an issue of greater importance in the near future. Managers planning to conduct restoration or enhancement projects in riparian areas near heavily traveled roads should consider building special safety measures into their projects, such as road underpasses or use of technologies to warn drivers of the presence of elk.

Low-mobility species are affected by roads at the time they are constructed. A direct loss of habitat for small mammals such as mice and rabbits, as well as a loss of animals in burrows, occurs where a road is constructed, but effects continue after the roads are in use. In Texas, narrow dirt roads are a barrier to movement by voles and cotton rats (Wilkins 1982), and in Ontario and Quebec, roads fragment populations of small mammals (Oxley et al. 1974). This negative impact is somewhat offset by the increase in food and cover plants in the disturbed soil. The practice of seeding forbs and grasses along paved highways to stabilize banks increases not only the amount of food available for some animals but also the probability of the animals being killed. Roads and their location must be considered in the planning process, especially if they are to be built near or in locations with sensitive habitats, such as meadows and riparian vegetation.

2.2.3 Fences

In the western United States, the thousands of miles of barbed wire fencing that cross the landscape to control the movement of livestock present a hazard, especially to antelope, which will not jump the fence. In addition, young deer and elk often will get caught trying to go through instead of over the wire. In Texas, fences that confined antelope herds to areas where livestock had depleted the forage resulted in 40% of the herd dying from toxic vegetation (Hailey et al. 1966). Also in Texas, antelope that could not get to quality forage during severe weather died from eating low-quality vegetation as a result of being trapped by sheep-proof fencing (Buechner 1950). The number of deaths caused by wire fences is uncertain, but the major impact occurs when fences block traditional migration routes of large animal species. As with roads, the location of fences must be considered in the planning process to find ways to reduce their impact on known migratory routes of large animals.

2.2.4 Weather

Disturbances caused by severe weather conditions such as hurricanes, tornadoes, heavy snow, rainstorms, and drought can affect large areas of forestland inhabited by wildlife. This could be especially disastrous if the forest contains endangered species. In one case, the red-cockaded woodpecker (*Picoides borealis*) lost nearly 100,000 ac of habitat and over 80% of its nest trees from the effects of Hurricane Hugo in 1989 on the Francis Marion National Forest in South Carolina (Wishard 1989). Prior to the hurricane, the forest had 477 family groups of woodpeckers, but 63% of the birds were killed by the hurricane (Hooper et al. 1990).

Malnutrition in white-tailed deer can be caused by snow depth and duration and is more severe when deer density is high (Severinghaus 1972; Bowyer et al. 1986). Young animals are affected by severe weather conditions, as are white-tailed deer fawns in Michigan; mortality can be as high as 50% (Verme 1977). During long periods of severe weather, many state agencies resort to supplemental feeding to sustain deer and elk herds through the winter months. This was the case in 2008 for elk in the Mount St. Helens Wildlife area when the Washington Department of Fish and Wildlife fed hay to more than 400 elk (Washington Department of Fish and Wildlife 2008). There is no doubt that weather has an effect on animal populations, and mortality losses have to be accounted for when state wildlife departments calculate their number of hunting permits.

2.3   DISEASES

A *disease* is an impairment of the normal state of a living animal or plant that affects the performance of the vital functions (Merriam-Webster 2008). Diseases usually affect individual animals by either killing them directly or so debilitating their physiology or behavior that they die from starvation or become easy prey for predators. Diseases are caused by bacteria (tularemia, botulism, Lyme); fungi (aspergillosis); viruses (foot-and-mouth disease, rabies); and toxic chemical substances, including pesticides. However, diseases are rarely a simple one-cause-effect relationship but are generally the sum of changes in the environment (Karstad 1979). Diseases transmitted from one animal to another by either direct or indirect contact are called *contagious*. Diseases capable of invading body tissue are termed *infectious,* and disease-producing agents are called *pathogens.*

2.3.1   DISEASE-PRODUCING AGENTS

Managers are concerned about diseases in wild animals because there are interactions between humans and wildlife, between livestock and wildlife, and between different species of wildlife. An example of a disease transmitted from wildlife to humans and to domestic animals from bites by skunks, foxes, and small mammals is rabies (*Rhabdovirus*). Rabies quickly overcomes the immunological system of an animal because the virus multiplies rapidly; such a disease is described as *virulent.*

One particularly feared disease is bubonic plague (caused by *Yersinia pestis*) or "black death." The fear arises from our knowledge of the large number of deaths that occurred in Europe in the 14th century (Tuchman 1978) and the fact that death from plague still occurs in the southwestern United States, where there are large rodent populations. For the period 1950–2007, there were 50 cases of plague reported in humans in Arizona (Arizona Department of Health Services 2007). The disease is not caused by rodents but by the bites of fleas and ticks (called *disease vectors*) inhabiting rodent burrows and dens.

Another disease transmitted by ticks is Lyme disease. Ticks, as intermediate hosts, get bacteria (*Borrelia burgdorferi*) from the blood of infected animals, particularly rodents, and then transmit the bacteria to humans or other animals when they suck blood (Davidson and Nettles 1988). The disease was formerly called Lyme arthritis and was named for the town of Old Lyme in Connecticut where it was first recognized. Lyme disease has become more prevalent in recent years and is often difficult to detect and treat. The second form of disease transmission, from livestock to wildlife and vice versa, led at least in one case to the destruction of a deer herd on the Stanislaus National Forest in the 1920s (Allen 1974). Foot-and-mouth disease (a virus) was being transmitted from deer to livestock in California and could only be controlled by reducing the deer herd. After eliminating over 34,000 deer, the disease was brought under control.

In 1993 in the southwestern United States, there were 10 deaths on the Navajo reservation as a result of a mysterious illness that struck 25 people in Arizona, New Mexico, and Colorado. The cause of the illness was traced to a *Hantavirus* infection that was spread from the droppings of rodents. Later, the rodents that were primarily responsible for carrying the disease were identified as deer mice in the genus *Peromyscus*. Deer mice are found in forests and woodlands across the United States, but all of the hantavirus cases were west of the Mississippi River. Because of favorable moisture conditions in the Southwest in 1993, large quantities of herbaceous vegetation were produced, which in turn produced abundant deer mice food—seeds and insects. As a result, there was a larger-than-normal increase in mice populations that could spread the virus.

Many diseases are transmitted from one wildlife species to another. An example is tularemia caused by the bacterium *Francisella tularensis*. The disease is transmitted by ticks and fleas and occurs in over 45 species of vertebrates but is present most often in rodents and rabbits (Davidson and Nettles 1988). Another form of disease arises out of naturally produced toxins. A good example is botulism, well known in the wildlife profession because it occurs with such frequency. The disease is caused by the bacterium *Clostridium botulinum,* which produces a toxin so powerful that when ingested by waterfowl it results in death. This bacterium typically develops in warm, stagnant water low in oxygen.

In recent years, humankind has added to the disease problem by introducing lead into the environment in hunting waterfowl. Waterfowl feed on lake and pond bottoms, where they pick up lead shot along with food items. The shot is retained in the gizzard, lead is absorbed into the bloodstream, and the waterfowl develop lead poisoning, which causes them to become emaciated, and their ability to walk or fly is impaired (Davidson and Nettles 1988). Raptors feeding on waterfowl with lead shot in their gizzard can also be affected with lead poisoning (Pattee and Hennes 1983).

Noninfectious diseases include such ailments as tumors (virus) and warts (virus). Also included in the disease category are parasites of roundworms, tapeworms, and flukes occurring in a host-parasite relationship. The immature parasite requires an intermediate host, in this case animals, for a "habitat." Readers who want a more in-depth understanding and background in diseases and parasites are referred to the work of Gilbert and Dodds (2001). Their coverage of the topics is current and well documented.

2.3.2 PESTICIDES, HERBICIDES, AND INSECTICIDES

Pesticides are chemical compounds used to control plants and animals, such as weeds, insects, rodents, and predatory animals. Likewise, compounds designed to kill insects are designated *insecticides,* and those designed to control vegetation are *herbicides.* For an account of the use of herbicides in the United States and how their once-widespread application directly and indirectly affected the ability of plants and animals to survive and reproduce, the book *Silent Spring* by Rachel Carson (1962) is an accurate presentation.

One of the best-known pesticides is DDT, a synthetic chlorinated hydrocarbon that was made available for public use in 1945 and was banned for most uses by the Environmental Protection Agency in 1972. It is the most persistent chlorinated hydrocarbon, followed by dieldrin, toxaphene, lindane, chlordane, heptachlor, and aldrin. Aldrin and dieldrin were banned in 1974, and heptachlor and chlordane were banned in 1975. Other pesticides include the organic phosphorus compounds of parathion and malathion used to kill parasites on plants and animals. Certain animals are more sensitive to pesticides than others. As a general rule, crustaceans (lobsters, shrimp, crabs), shellfish (clams and oysters), and fish are most sensitive, followed by amphibians (frogs, salamanders), reptiles, birds, and mammals (U.S. Department of Interior 1966).

Pesticides may kill animals directly, but most exert indirect effects through the food chain either by depletion of a food supply or reduction of reproductive capacity after an animal consumes contaminated food species. Fish and birds are good indicators of pesticide contamination. Birds feed on insects that may have been sprayed, and fish come into contact with runoff water containing the contaminant. Pesticides not only affect wildlife but also humans through the food chain or by direct contact. Most people in North America have measurable amounts of pesticides in their bodies (Council on Environmental Quality 1981).

2.4 PREDATORS

A *predator* is an animal that kills and feeds on other animals. Hawks eat mice, cougars eat deer, and foxes eat rabbits. Because of this "kill-and-eat" trait, probably no other topic in wildlife management elicits as much emotion and debate as the subject of predators, particularly as they affect animals on which humans place great value for hunting or viewing, such as deer, quail, squirrels, and rabbits.

2.4.1 THE LEOPOLD REPORT

A background statement on wildlife management, prepared by an advisory board for a report, known as the Leopold report, summarized the attitudes about predators that once existed in the United States and how these attitudes are changing:

In a frontier community, animal life is cheap and held in low esteem. Thus it was that a frontiersman would shoot a bison for its tongue or an eagle for amusement. In America we inherited a particularly prejudiced and unsympathetic view of animals that may at times be dangerous or troublesome. From the days of the mountain men through the period of conquest and settlement of the West, incessant war was waged against the wolf, grizzly, cougar, and the lowly coyote, and even today in the remaining backwoods the maxim persists that the only good varmint is a dead one. But times and social values change. As our culture became more sophisticated and more urbanized, wild animals began to assume recreational significance at which the pioneer would have scoffed. Americans by the millions swarm out of the cities on vacations seeking a refreshing taste of the wilderness, of which animal life is the living manifestation. Some come to hunt, others to look or to photograph. Recognition of this reappraisal of animal value is manifest in the myriad of restrictive laws and regulations that now protect nearly all kinds of animals from capricious destruction. Only some of the predators and troublesome rodents and birds remain unprotected by law or public conscience. (Leopold et al. 1964, pp. 27–49)

The Leopold report is a good synopsis of the "sign of the times" when predator control was a major issue for state game and fish departments and federal land management agencies. The report reflects the attitudes that once existed but with an outcome that all animals have value; however, at different times some have more value than others (Axiom 7).

Since the report was written, additional legislation protecting animals classed as predators has been enacted. These laws emphasize the role of animals in maintaining ecological diversity, social and scientific value, and ecosystem functioning. Over millions of years, predators have evolved together with their prey. If predators destroy their prey completely, then the predator will die. So, a delicate balance between prey and predator has evolved. In ecological systems, predators are believed to perform the function of removing excess numbers and culling the prey population by removing weak, diseased, or abnormal individuals.

2.4.2   Lotka-Volterra Equations

The first theoretical work done on predator–prey relations was that of Lotka (1925) and Volterra (1931), who developed the basic equations in the 1920s (Chapman 1931). The final result of these equations is a cycle in which as a prey species population increases, the predator population increases but with a lag. These increases result in a reduction in prey population lagged by a reduction in predator population, and so the cycle repeats. The Lotka-Volterra equations do not consider all the interactions taking place in a natural ecosystem, such as populations of other prey species (buffer species) that are available when the primary prey species is reduced by the predator. So, as a practical tool for field biologists, the equations are of little value. The equations are, however, of historical value because they generated considerable interest among scientists studying animal cycles, resulting in a better understanding of the predator–prey relationship.

One of the first studies on predator–prey relationships was in Iowa on mink predation of muskrats (Errington 1967). The results generally confirmed the findings that when a muskrat population was high, predation by mink was also high, but predation was not even across all age classes. Errington also discovered that social interactions within muskrat populations contributed to limiting their populations. In Errington's study, the muskrats selected were those suffering from disease and starvation, so the predation was compensatory, not additive. This simply means that the predator replaces disease and starvation as the decimating factor. If predation were additive, then the loss of animals would be added to that caused by disease and starvation. The idea that predators take only the sick and weak is intuitive and not well documented and oversimplifies the predator–prey relationship.

2.4.3 GENERAL PROPOSITIONS

Some general propositions of predator–prey relationships that can be formulated from ecological studies are as follows:

- The density of a predator population is almost always less than that of its prey.
- Predators feed on a variety of prey when available and not just a single species.
- Reproduction rates of prey species are far greater than of predator species.
- The prey species is almost always smaller in size than the predator, but when small predators prey on larger species, it is usually on the young.

One study of predator control often cited as the reason for an increase in deer populations is centered on the Kaibab Plateau in Arizona. Before predator control was begun in 1906, the North Kaibab had an estimated 4,000 deer (Rasmussen 1941). From 1906 to 1923, cougars, wolves, coyotes, and bobcats were removed by hunters and trappers. Twenty years later, just before a die-off, the deer population was estimated to be 100,000 on the plateau; however, the accuracy of this estimate is questionable. The results are confounded by the fact that 200,000 sheep and 20,000 cattle were removed from the study area during the same period (Caughley 1970).

An interesting idea that developed from predator–prey studies is that predators can maintain species diversity (Spight 1967). Diversity is enhanced because predators keep abundant species in check so that they cannot outcompete less-abundant species for food, shelter, and other necessities. The thought to keep in mind is that, as in the Lotka-Volterra equations, a strict predator–prey relationship seldom provides all the answers to the question. Predators are an important component of the environment of an animal, but the effects of predation on a population must be considered in connection with the other environmental factors.

All living things die and are recycled as part of the flow of energy through the life community or stated another way, a creature must feed and sooner or later it will be fed upon. (Allen 1979)

When considering predator–prey interactions, information is needed on five factors before the relationships can be understood (Leopold 1933). These factors are

- Density of the prey population
- Density of the predator population
- Prey characteristics, such as reactions to predators and nutritional condition
- Density and quality of alternate foods available to the predator
- Predator characteristics, such as its means of attack and food preferences

The controversy about predators and their place in ecosystem management continues today, and it is easy to get the public enraged on two sides: (1) those who support predator control and will oppose the reintroductions of any carnivore, such as the gray wolf, into its former range and (2) those who do not want predators controlled but will support reintroduction of species that have been eliminated from their historic range.

2.5   HUMANS

Early humans were part of the natural environment and had little impact on ecosystems until they domesticated animals, acquired tools and weapons, learned to use fire, and were able to affect animal populations by hunting. The effect of the Plains Indians on the bison populations was minimal compared to the effects of the railroads, guns, and livestock grazing. Today, however, it is obvious that resources are being used more extensively and intensively by expanding human populations. In many parts of the world, humans are now in conflict with and will increasingly be in conflict with other animal species for the use of the limited resources of the earth. An understanding of the conflicts between humans and wildlife is one of the keys to the well-being of both.

There is no doubt that humans directly (legal and illegal hunting) and indirectly (habitat destruction) affect the reproduction and survival of a large number of animals. Also, there is something about the presence of humans that causes undomesticated animals to flee or to maintain their distance. Animals can become adapted to human presence after several generations of being in a local population where the interaction occurs, but these interactions often lead to problems as animals such as deer first become tame and then a nuisance. Humans as hunters belong in the centrum of an animal, but humans as destroyers of habitat properly belong in the web. While hunting, both legal and illegal, affects animals directly, we can assume that legal hunting, if part of a continuing system, is under control, but illegal hunting is a different situation.

2.5.1   HABITAT DESTRUCTION

Habitat destruction by humans undoubtedly poses a severe indirect impact on wildlife. But, care must be taken in describing destruction because the loss of habitat for one animal is often a gain for another. So, destruction must be considered in terms of the species affected. However, there are three forms of habitat destruction that adversely affect all wild species:

- Urbanization
- An increase in clean agriculture (removal of all plant material)
- Soil loss

Conversion of forestland to cropland and urban uses such as roads and reservoirs is projected to increase, with the most significant being conversion of bottomland hardwoods to cropland in the lower Mississippi River area (National Academy of Science 1982). In 1937, there were an estimated 11.8 million acres of bottomland hardwoods in that region; about 5.2 million acres remained in forest in 1978. Efforts to restore bottomland forests in the Mississippi alluvial valley have been undertaken by the Lower Mississippi Valley Joint Venture Forest Resources Conservation Working Group (Wilson et al. 2007). The group recommends landscape level areas in a matrix of large blocks of contiguous forest for restoration and maintenance of forests and wildlife habitat. They further recommend using restoration prescriptions through adaptive management, and that forest management decisions should be evaluated with regard to habitat conditions and wildlife response.

Large losses in habitat for big game species have been documented in the West. For example, in Wyoming, a state of wide open spaces, it is estimated that from 1967 to 1977 approximately 150 mi^2 of forest land were urbanized and lost as deer and elk habitat (Rippe and Rayburn 1981). In the Southwest, riparian vegetation consisting of large deciduous trees is dwindling due to the activities of woodcutters, recreationists, grazing, and an increase in home construction. Riparian areas are now recognized as important and critical habitat for many species, but the loss of riparian habitat has not been documented adequately due to lack of long-term inventory data.

As human populations increase, there must be a corresponding increase in living, work, and commercial space as well as in food-growing areas, railroads, and highways for transport. Since the land base is fixed, such increase must necessarily be appropriated from land used by wildlife species. The loss of habitat to urbanization will inevitably proceed as long as human populations continue to increase. The direct effect from habitat destruction is a loss of resources, but humans remain indirectly the major cause of the effect on wildlife.

2.5.2 Illegal Hunting

The impact of illegal hunting (poaching) on populations of wildlife species is difficult to determine. Market hunting as it existed in the 1800s to the 1930s is no longer a threat. All U.S. and Mexican states and Canadian provinces to varying degrees now have conservation or other law officers in place to enforce hunting and fishing regulations. In some countries, special antipoaching units exist, and the public, particularly in the United States and Canada, is encouraged to report violations.

Approximately 10,000 white-tailed deer are illegally killed each year in Missouri (Witter 1980). Studies in Idaho have provided estimates that in 1967 out-of-season deer kills were 8,000 at a time when the legal harvest was 66,000 mule and white-tailed deer (Vilkitis 1968). On the Kaibab Plateau in Arizona, a mortality study using radio-collared mule deer indicated that poaching could account for as much as 10% of the loss of instrumented deer (McCulloch and Brown 1986). The evidence, while not well quantified, does confirm that poaching continues to be a problem in some areas. The problem is aggravated by evidence that local residents often tolerate illegal hunting (Decker et al. 1981).

2.6 BIOLOGY

The innate biological factors leading to the development of a population include reproductive and behavioral traits specific to a species. These species characteristics are contained in the general category of genetics in the biology component of the centrum. At present, managers may not be able to change the innate characteristics of a species, but genetic engineering provides possibilities for the future. Genetic factors determine how a species responds to its environment, and in this respect they are directly acting.

2.6.1 REPRODUCTION

Breeding season, age of sexual maturity, gestation (incubation), longevity, and fecundity are all elements of a the reproductive characteristics of a species. These characteristics are innate and have a wide range of values. *Breeding season* refers to the time of year mating can take place and is variable among species. Breeding season within species occurs at the same time, so that there is maximum opportunity for the species to perpetuate itself. Time of breeding is controlled by day length (photoperiod), but breeding is also influenced by temperature and the general physical condition of an individual.

Breeding seasons within the geographical range of a species may differ; in West Virginia, the white-tailed deer breeds in November, but in Arizona, breeding in some areas occurs in January. Animals that produce many young (mice, rats, etc.) generally breed in the spring, and the young are born in the spring or summer when the environmental conditions are most favorable for their survival. Animals, such as deer, that produce only one or two young breed in the fall with the young born the following spring or summer.

Two general types of breeding behavior are evident within a breeding season. A *monogamous* species is one in which a single male and female pair for the breeding season and maintain strong social bonds. This contrasts to *polygamy,* by which one male or female breeds with more than one female or male. In polygamous species, a social bond forms but is not as strong as in monogamous species. One type of polygamy is *promiscuity,* in which one male breeds with many females or vice versa, but no social bond is formed. For example, the bald eagle is a monogamous species, elk are polygamous, and cottontails are promiscuous.

Age at sexual maturity determines whether an animal can breed within a breeding season, although poor nutrition may delay onset of breeding. Small mammals generally breed when they first become sexually active. Deer, in contrast, born in the spring generally breed the fall of the next year (1.5 yr of age) if they are in good physical condition. The length of time a female is pregnant defines the *gestation* period. Species with short gestation periods are capable of producing the greatest number of young, and some species can breed more than once in a breeding season. The gestation period in mammals is analogous to the incubation period in birds.

The longevity (age) of an animal species determines how many times and how many young a female produces in a lifetime. Longevity is of two types: physiological and ecological. *Physiological longevity* is the maximum life span of a species under optimum conditions. Some examples of animal ages acquired from zoo records, which should approach physiological longevity, are presented in Table 2.1. *Ecological longevity* is the realized longevity of a species in the wild. Long-lived animals generally produce the fewest number of young per breeding season. Contrast the number of young per litter, number of litters per breeding season, minimum breeding age, and longevity of the cottontail rabbit (low mobility) with that of white-tailed deer (high mobility) (Table 2.2). The number of offspring produced per female of a species is *fecundity* (fertility). The two categories of low and high fecundity are known as *r-selection* and *K-selection,* respectively (MacArthur and Wilson 1967).

TABLE 2.1
Physiological Longevity from Zoo Records

Species	Years
Deer mouse	4
Cottontail	10
Mountain lion	20
Cottonmouth	21
Elk	22
Bullfrog	30
Grizzly bear	32
Great horned owl	68

Source: McWhirter, N. *Guiness Book of World Records,* Sterling, New York, 1981.

TABLE 2.2
Contrasting Characteristics between the Cottontail Rabbit (Low Mobility) and White-Tailed Deer (High Mobility)

Factor	Cottontail Rabbit	White-Tailed Deer
Young/litter	4–6	1–2
Litters/year	2–4	1
Breeding age	3 mo	1.5 yr
Ecological longevity	3 yr	8–10 yr

The former refers to a high intrinsic rate of increase (r), and the latter refers to the ability to maintain a population that fluctuates around a carrying capacity (K).

2.6.2 BEHAVIOR

Territories and dominance hierarchies are elements in the social behavior of some animal species, and both appear to provide survival value, although this is hard to document. A question concerning population biologists at present is whether these two factors limit animal populations, as has been suggested by Wynne-Edwards (1962). In spite of the difficulties of sorting out behavioral problems associated with territories and hierarchies, there are some advantages for animal species:

- Limited resources are preserved for the small number of individuals in the defended area.
- Males are ensured of reproducing with one or more females.
- The defending males are the dominant males of the species; therefore, genes of the strongest are passed on to the young.

2.6.3 TERRITORY

The behavior of animals affects their interactions with members of the same species, particularly during the breeding season. Studies of birds led to the idea of a *territory* as a defended area that is part of the home range of an animal (Howard 1920; Nice 1941). More recent definitions of territory explained that it is an exclusive area (Schoener 1968) or any spacing of animals that is more even than random (Davies 1978). Different definitions have arisen because territory is better defined for some life forms, such as birds, than it is for others, such as mammals. However, mammals do have territories, but they are more difficult to detect and are not always associated with breeding, as has been demonstrated for cougars in Idaho (Hornocker 1969). In this study, cougars had winter territories where all individuals shied away from each other, but they often crossed territories.

Because territory is an expression of space, there has to be some way for the territorial animal to mark and identify the space. In mammals, this is done with fecal material or urine, but in birds territories are typically defended by displays (an act of aggression) or calls to mark a boundary; however, physical violence is rare because one individual usually breaks off the encounter. Territories can be horizontal, as is the case of some mammals, or vertical for birds. Some large carnivores, such as the wolf, maintain a group territory and defend it against intruders (Mech et al. 1971).

There is evidence that the nature of territories changes as animal density changes (Brown 1969). At low animal densities, territories are located in preferred habitat, while at high densities some individuals are forced into less-suitable or marginal habitat. Animals at low density defend larger territories, but when density is high the territory is smaller (Kendeigh 1941). Closely related to territory is *hierarchical dominance,* which develops in social animals that form groups. These types of hierarchies are found in amphibians, birds, fishes, invertebrates, and mammals (Wilson

1975). Hierarchical dominance establishes a *pecking order* within the social group. This order ensures the dominant males of mating opportunities, such as in wolves (Fox 1980), and of food, as in white-tailed deer (Severinghaus 1972). In the case of white-tailed deer, the adult bucks dominate does, while does dominate fawns in the order of feeding in winter deer yards.

2.6.4   ECOLOGICAL NICHE

Defining the role of animals in the functioning of ecosystems sometimes leads to complex theories that are difficult to explain and understand. These difficulties often result in confusing usages among practitioners. One such theory is of an *ecological niche* (Grinnell 1943). The theory of a niche derives from the way an animal uses its habitat and the special adaptations either in physical structure or behavioral characteristics that facilitate this use. For example, squirrels climb trees, deer browse woody plants, and some rodents create burrows.

The niche of each species is developed through evolution. Each niche therefore is unique and distinct from all others. This in turn leads to an ecological principle that holds that no two species can occupy precisely the same niche and is commonly stated as the principle of *competitive exclusion*. Criteria defining niches are hierarchical and often overlap at higher levels. One consequence of the principle of competitive exclusion is that if two species compete at the niche level, then one species ultimately displaces the other or evolves to develop a new niche. Probably the easiest way to understand a niche is to examine how a particular species uses a resource also used by another species in the same geographical area.

For example, the tassel-eared squirrel eats the inner bark of ponderosa pine clipped from twigs. The porcupine also eats inner bark, but from the upper trunk and limbs. Although the two species use the same food resource, they differ in where they eat and in the size of stems used. The squirrel and porcupine both feed in trees. At the next-lower level, they feed on bark, so these two species have at least two levels of overlap in their requirements, but as the levels are more narrowly defined, the two species separate in their use of resources. Criteria that define niches are hierarchical and often overlap at low levels of resolution. A major point to emphasize is that a niche is a functional characteristic of a species and is not just a physical location. The number of niches in a habitat is also a function of primary productivity. Production increases from a minimum at polar regions to a maximum at the equator; therefore, more niches occur in equatorial forests than in forests in northern latitudes.

2.6.5   POPULATION CHARACTERISTICS

Environmental factors focus on what influences the ability of an individual animal to survive and reproduce. Individual animals, however, do not reproduce alone but are always part of a community organization. Therefore, all wildlife species function as a population within a habitat. Populations can only increase or decrease as a result of four factors: natalite, mortality, immigration, and emigration. Such a simple statement does not describe changes in the population over time, which is the concern of *population dynamics*.

The study of populations is fascinating but can be complicated. A detailed population analysis requires mathematical tools, including calculus, which deals with rates of change and limits. However, in this chapter population dynamics is presented largely on an intuitive basis, depending only slightly on advanced mathematics. A population (P) is a social, interbreeding group of animals occupying space and characterized by such qualities as natality (birth rate), mortality (death rate), and density, which are not characteristic of individuals.

Fecundity of individuals underlies the natality of a population. *Natality* (N) is the number of new individuals added per year through birth in a population and is expressed as either a crude birth rate or a specific birth rate. A *crude birth rate* describes the number of individuals born per unit of population, for example, 10 births per 100 individuals. A *specific birth rate* adds a factor such as age, for example, the number of young produced per year by 3-yr-old females.

The number of individuals in a population dying during a given period of time, usually a year, is the *mortality* (M) of that population. Mortality is expressed as a rate per unit, such as number per 100, 1,000, and so on, and can be given for the total population or specific age classes. *Immigration* (I) and *emigration* (E) refer to animals that join a group from an outside breeding population or leave a breeding population to join another, respectively. If natality and immigration are higher than mortality and emigration, then a population will increase; otherwise, it will decrease. In any case, the number of animals present per unit of area is reflected in population density P_D as

$$P_D = (I + N) - (E + M)$$

Two types of density can be defined, and it is important to distinguish between them. *Crude density* is the number of animals per unit area regardless of what resources and other factors the area contains. An example is the number of deer per square mile even though part of the area may include a town and roads that are not deer habitat. On the other hand, *ecological density* includes only those areas considered suitable habitat for a particular species. An example is the number of deer per acre or section in a given forest type (oak-hickory, ponderosa pine).

The density of a population is given in number of animals per unit of area, but there is no simple measure for distribution. Three different areas may have the same population density, but one may be homogeneously distributed, another randomly distributed, and the third distributed in a randomly clumped manner (Figure 2.1). The pattern of animal distribution is important because pattern influences the manager's choice of inventory technique.

Populations are of two types: a *local population,* which is a group of interbreeding animals restricted to the use of a particular area, or a set of local populations that interact as a larger unit, called a *metapopulation* (Smith and Smith 2009; Primack 2006). Over a period of time, a local interbreeding group may adjust to environmental conditions to a degree that they genetically could become a subspecies, as is the case of the Kaibab squirrel on the north rim of the Grand Canyon. A metapopulation seems to be the same as a *natural population,* which was described by Andrewartha and Birch (1984) as the sum of all local populations not separated by a barrier.

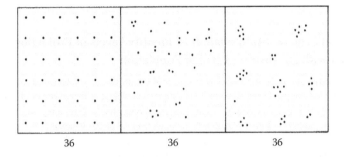

FIGURE 2.1   Populations with the same density but three different distributions.

Populations vary greatly in the size of area they use, depending on animal density, mobility, and availability of resources. A local population may exist on the side of a mountain or on several mountains but use the same place consistently over time. Local populations probably arise out of the imprinting of young to the area, thereby enhancing survival and reproduction. Occasionally, one or more animals in a local population mates with an individual from another local population, so there is exchange of genetic material. Individuals in a local population adapt to the local environmental conditions, and genetic exchange between individuals makes the local population the basic unit in evolution.

In some cases, a metapopulation may encompass large areas, such as the entire geographical range of a species, but as long as there are no barriers to breeding between local populations, the definition holds. The metapopulation may display a steady-state density without large yearly fluctuations despite the fact that local populations may be high in one area and low in another. Thus, it is possible for a local population to go to extinction while the metapopulation continues to survive and reproduce, and new local populations could be established from sink populations. The characteristics and usefulness of the metapopulation concept is in the formative stage and will continue to be a topic for discussion as research provides additional information and insight.

The National Forest Management Act (U.S. Congress 1976) introduced the term *minimum viable population*, but no clear definition was given. Biologically, a *viable population* is the number of individuals sharing a common gene pool and able to reproduce and maintain the population from one generation to the next as in a local population. Viable population concepts are only beginning to appear in the literature, so there will undoubtedly be a time of testing, evaluating, and changing definitions and procedures for measurement before the concept is fully mature. The question that must be answered is, What is the number of local populations necessary to maintain a natural viable population? The question is important for managing wildlife because of the trade-offs that may be necessary to maintain viable populations of other species in the same area.

What options do managers presently possess to guide them in the planning for and management of a viable population of a given species? Two seemingly appropriate techniques are currently in use: estimated protection levels and risk analysis. The

TABLE 2.3

Hierarchical Scheme (Abbreviated) to Identify Levels of Protection Associated with an Abstract Level of Population

Level 1. Individual survival—low likelihood of survival for any time beyond a few decades. Several individuals or pairs isolated on the area with no interchange with the species off the area.

Level 3. Family resilience—survival for a half century or longer likely. Several reproductive or social groups partially to fully isolated in the area, with a total of 20–50 adults.

Level 5. Short-term adaptability—continued existence beyond a century is likely. A well-distributed local population with an effective biological population (Ne) in the low to middle hundreds.

Level 7. Long-term adaptability—continued existence for many centuries likely. A well-distributed population with an Ne approaching 1,000.

Level 9. Evolutionary viability—continued existence on the order of millennia is likely. Populations are fully capable of evolutionary change. Well-distributed, local populations, parts of a biological population with an Ne that exceeds the low thousands.

Source: Schonewald-Cox, C. M. Guidelines to Management: A Beginning Attempt. In C. M. Schonewald-Cox, S. M. Chambers, B. MacBryde, et al. (Eds.), *Genetics and conservation, a reference for managing wild animal and plant populations*, pp. 414–445. Benjamin-Cummings, Menlo Park, CA, 1983.

former is a scheme to identify hierarchical levels of protection at abstract levels of populations. At present, nine levels of protection have been established; five of them are described in Table 2.3. One of the major problems in assigning a population level to one of the nine categories is making a reasonable estimate of the population density. Clearly, when populations are high and well distributed, the probability of maintaining a viable population is high. The situation is reversed in the case of low, dispersed populations, but much uncertainty arises in assigning species populations to some of the intermediate levels.

When uncertainties in making a decision occur, a risk analysis may help in making management recommendations. Risk analysis deals with two types of uncertainty: scientific and decision making (Marcot 1986). Scientific uncertainty centers on the lack of biological information about a species and the consequent inability to adequately predict either long-term natural or short-term catastrophic events. Uncertainty in decision-making results from the use of inadequate or inaccurate information. In dealing with the uncertainties, both scientific and decision making, risk analysis involves estimating probabilities of various events, estimating values of outcomes of various possible decisions, and finally applying the resulting probabilities of outcomes to estimate a range of results for a variety of possible decisions. The use of a risk analysis reduces some of the uncertainty in making a decision. However, given the intractable uncertainty of the future, it is clear that science cannot furnish wildlife managers with an absolute deterministic scenario of events. "A precise forecast of the future is excluded if we can have only an inaccurate measurement of present circumstances" (Weaver 1975, p. 300).

2.6.6 Uninhibited Growth Model

How does the mating of several individuals produce a population? To provide a theoretical answer to this question and to demonstrate how a population might grow, a hypothetical animal population can be developed conceptually by imagining the outcome of placing one male and one female with a physiological longevity of 10 years in an enclosure of 1 mi². The enclosure is stipulated to be free of predators, hazards, and diseases and possesses the optimum resources for the species. Further, it is stipulated that two young are produced in June each year (female was bred the previous fall), with a 50:50 sex ratio at birth and one breeding season per year. At the end of the first year, the enclosure will contain four animals consisting of two males and two females (Table 2.4). In the second year, both females breed and produce two young each for a total of four new animals.

At the end of 10 years, there are 2,048 animals in the enclosure. A population increasing in this manner develops a characteristic J shape when plotted over time (Figure 2.2) and represents the *uninhibited growth model.* Uninhibited or *exponential growth* results when animals are introduced into unoccupied habitat (designated as a *pioneering population*) or under laboratory conditions. Individuals born into a population at the same time are referred to as a *cohort.* In our hypothetical animal population, a cohort turnover occurs at 11 years when the first pair introduced into the enclosure dies. Notice in this animal population model that, although there is a constant growth rate (two offspring for each mature female), the number of individuals added each year changes dramatically. Adding 10% to a population when it is low $(0.10 \times 100 = 10)$ is different than adding 10% when it is high $(0.10 \times 1,000 = 100)$

TABLE 2.4

Uninhibited Growth for a Hypothetical Population with a 50:50 Sex Ratio, One Breeding Season per Year, and a Physiological Longevity of 10 Yr

	Adult		Young		Subtotal		
End of Year	M	F	M	F	M	F	Total
1	1	1	1	1	2	2	4
2	2	2	2	2	4	4	8
3	4	4	4	4	8	8	16
4	8	8	8	8	16	16	32
5	16	16	16	16	32	32	64
6	32	32	32	32	64	64	128
7	64	64	64	64	128	128	256
8	128	128	128	128	256	256	512
9	256	256	256	256	512	512	1,024
10	512	512	512	512	1,024	1,024	2,048

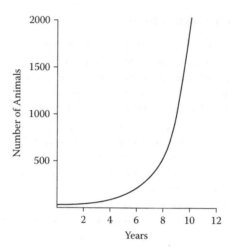

FIGURE 2.2   Uninhibited growth model.

and in the same environment may be catastrophic. This factor leads directly to a consideration of rates of population change.

2.6.7   RATES OF POPULATION CHANGE

Net population increases are a function of the number of young born to females of specific ages. The *net reproductive rate* (R) is the average number of females born per female in each age group. The value of R is included in fecundity tables, but it is difficult to determine and is almost never available to wildlife managers. However, an approximation of R is provided by the finite rate of change. The total yearly change in a population can be expressed as a growth rate in the ratio of population numbers from one year to the next. This rate is the *finite rate* (Greek letter lambda, λ) or the rate of change of population size. Finite rates are observed or actual rates obtained from population estimates. For example, using information from Table 2.4, the finite rate of change in population from Year 2 to Year 3 is given in Population Equation 2.1:

$$\lambda = P_3/P_2 = 16/8 = 2 \tag{2.1}$$

where P_3 is the population at Year 3, and P_2 is the population at Year 2.

The base population of 8 changed by a factor of 2 (100%) in 1 yr. Subsequent years can be calculated by the general equation

$$P_{t+1} = \lambda \times P_t$$

where P_t is the population for a given year, and t + 1 is the population for the next year.

For Year 5, the calculation is

$$P_5 = \lambda \times P_4 = (2.0 \times 32) = 64$$

Notice that the rate of change in our hypothetical population is constant between years. A population increasing at a constant growth rate follows a sequence of increasing power of λ such that (Population Equation 2.2)

$$P_t = P_0(\lambda)^t \tag{2.2}$$

where P_0 is the initial population, and P_t is the population in Year t.

This equation assumes that the species will breed only once during the breeding season and does not have overlapping generations. For example, the animal population in Year 5 with a λ of 2 is

$$P_5 = 2 \times 2^5 = 64$$

A field example of exponential growth occurred when 2 male and 6 female pheasants were introduced on Protection Island in 1937 (Einarsen 1945). The population increased from 8 to 2,000 pheasants in 6 yr. Other examples of uninhibited growth include deer introduced into the George Reserve (McCullough 1979) and reindeer on St. Matthew Island (Klein 1968). To simplify rates, it is convenient to use the delta (Δ) notation to express a change in quantity to distinguish finite changes from instantaneous changes (d) used in derivatives in calculus. A change in quantity can occur in either the time or the quantity factor. Another way to express the rate of change of a population is to determine, via Population Equation 2.3, an average rate per individual (r_0) per unit of time as

$$r_0 = \frac{\Delta P}{P_t(\Delta t)} \tag{2.3}$$

where r_0 is the rate of change per individual per unit of time or specific growth rate, ΔP is the change in population, Δt is the change in time, and P_t is the population in Year t.

The rate of change for our animal population changing at a constant yearly rate is (select any two years, such as Year 2 to Year 3) (Table 2.4):

$$r_0 = \frac{P_3 - P_2}{P_2(t_3 - t_2)} = \frac{16 - 8}{8(3 - 2)} = 1$$

The interpretation is that every animal (male and female) in existence in Year 2 will account for one animal to be added to the population in Year 3. The rate of population change per unit of time can be rewritten as in Population Equation 2.4:

$$\Delta P/\Delta t = r_0 P_t \tag{2.4}$$

By knowing the rate of change (r_0) per individual, we can grow a population from an initial number over time. For example, the population in Year 3 is

$$P_3 = P_2 + r_0P_2 = 8 + (1 \times 8) = 16$$

The finite rate of increase (λ) may be expressed in exponential form as follows:

$$\lambda = e^r$$

where e is 2.71828 (the base of natural logs), and r (natural log λ) is the intrinsic rate of increase.

In this equation, r is an exponent to which e is raised to equal λ and is a rate of increase that can be used in population equations. For our hypothetical population, λ is 2; therefore, the value of r is 0.693147 (the natural log of 2). Finite rates are always positive and have a range from zero to infinity, but when λ is greater than 1, the population will increase, and when λ is less than 1, the population will decrease (Table 2.5). In population analysis, r is used instead of λ (Caughley 1977) because

- It is centered at zero, and an increase is positive (+) while a decrease is negative (−).
- It converts easily from one time unit to another. Thus, to convert a yearly rate to a daily rate, divide by 365.
- It can be used to calculate the doubling time of a population by the formula

$$0.6931/r = \text{doubling time } (Dt)$$

where 0.6931 is the natural log of 2.

For example, from Year 3–4, $\lambda = 2$, $r = 0.6931$:

$$Dt = 0.6931/0.6931 = 1$$

Whenever the percentage of animals in each age class of a population remains constant over time, the result is a stable age distribution. From the uninhibited growth data (Table 2.4) of our hypothetical animal population, a table (Table 2.6) can be developed that shows the population by age class. A population that is increasing

---

TABLE 2.5

The Relationship between Percentage Change, Lambda λ, and Exponential Rate r

P_y	P_{y+1}	λ	% Change	r
100	50	0.50	−50	−0.693
100	100	1.00	0	0.000
100	150	1.50	+50	+0.406
100	200	2.00	100	+0.693
100	500	5.00	400	+1.6094

TABLE 2.6

Stable Age Distribution for a Hypothetical Animal Population from Table 2.4

End of Year	Number at Age								Total Population	% of Age Zero in Population[a]
	<1	1	2	3	4	5	6	7	8		
1	2	2	0	0	0	0	0	0	0	4	50
2	4	2	2	0	0	0	0	0	0	8	50
3	8	4	2	2	0	0	0	0	0	16	50
4	16	8	4	2	2	0	0	0	0	32	50
5	32	16	8	4	2	2	0	0	0	64	50
6	64	32	16	8	4	2	2	0	0	128	50

a Age class < 1 divided by total population in a given year.

at a fixed rate will reach an age distribution that is unchanging. This is called the *stable age distribution*. When a stable age distribution is attained, the population will increase according to Population Equation 2.5:

$$P_t = P_0 e^{(rt)} \qquad (2.5)$$

where P_t is the population at time t, P_0 is the initial population, e is 2.71828 (base of natural logs), r is the intrinsic rate of change, and t is the time in years.

Because the value r is the rate of change per individual per unit of time, it accounts for not only the difference between births and deaths but also emigration and immigration. This rate can be used in very short intervals of weeks and days. Using Equation 2.5, our hypothetical population for Year 5 is

$$P_5 = P_0 e^{r(5)}$$

$$P_5 = 2 \times 2.71828^{(0.693147 \times 5)} = 64$$

The same equation can be used to reconstruct a population growth rate if the initial and ending population numbers are known. For example, if a pioneering population of pheasants increased from 8 to 2,000 in 6 yr (Einarsen 1945) when introduced into unoccupied habitat, then what were λ and r? The calculations are as follows:

$$P_t = P_0 e^{(rt)}$$

$$P_t/P_0 = e^{rt} = \lambda$$

$$2,000/8 = e^{rt} = 250$$

$$r = \text{natural log of } \lambda = 5.5215$$

$$r = 5.5215/6 = 0.9203 \text{ (for 6 yr)}$$

TABLE 2.7

Effects of a Change in Litter Size on Population Growth (50:50 Sex Ratio, 1 Litter per Year, No Mortality)

	Year	
Average Litter Size	1	5
1	3	15
2	4	64
5	7	1,052
10	12	15,552

Substituting figures into Equation 2.5,

$$P_6 = 8 \times e^{(0.9203 \times 6)} = 2,000$$

Since the world is not overrun with animals, as the uninhibited model seems to imply, there must be something controlling populations so that numbers do not increase without bounds. But, before approaching this question, several factors that condition population growth must be noted. First, the illustration used to derive the "J-shaped" curve assumes that females breed when 1 yr old. If instead the females in the hypothetical animal population first breed at 2 yr, the increase in numbers is dramatically reduced, so the shape of the growth curve changes to reflect a slower growth rate. Further, the example assumed a 50:50 sex ratio at birth, which seems to be normal for most vertebrate species (Caughley 1977). But, what if the number of young born is greater or less than two? Table 2.7 shows how annual production changes with a change in litter size (note the assumed 50:50 sex ratio).

2.6.8 INHIBITED GROWTH MODEL

Obviously, populations cannot continue to expand indefinitely. Sometime during the uninhibited growth of a new population, the rate of increase r must start to decline. This change of rate results from the interactions and consequences of the factors contained in Conceptual Equation 1.1. After these factors become fully effective, a point is reached at which the population levels off and reaches an equilibrium with its environment (Figure 2.3). Exponential growth moving to a steady state over time is reflected by an "S-shaped" curve. The upper part of the curve, however, can take several shapes in addition to the one shown in the figure. The inhibition of the growth of a population can be dealt with mathematically by inserting a variable in the uninhibited growth formula (Equation 2.2) to reduce the value of r as the density increases. This expression is

$$(K - P_t)/K$$

where K is the carrying capacity (resources), and P_t is the population at time t. This shows the same S-shaped sigmoid (logistic) curve as in Figure 2.3.

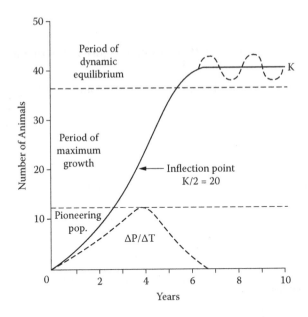

FIGURE 2.3 Inhibited growth model for a hypothetical population.

If the enclosure containing the hypothetical population possesses resources sufficient to support only 40 animals, the closer the population approaches 40, the closer the value of the expression approaches and eventually equals zero as a limit. At carrying capacity K, population growth is zero. To calculate the annual population for Column 2 of the logistic growth curve (Table 2.8), the formula in Population Equation 2.6 is used:

$$\frac{\Delta P}{\Delta t} = r_0 P \frac{K - P_t}{K} \tag{2.6}$$

For example, the logistic growth (yearly addition) from Year 6 to Year 7 is

$$r_0 = \frac{\Delta P}{P(\Delta/t)} = \frac{256 - 128}{128 \times 1} = 1$$

Substituting numbers into Equation 2.6:

$$1 \times 39 \frac{40 - 39}{40} = 1$$

One of the interesting features of the logistic curve is that maximum production occurs at about $K/2$, which approximates the *inflection point*, where the population is still increasing but at a decreasing rate. For the hypothetical population in

TABLE 2.8

Inhibited Growth of the Population in Figure 2.3 with a *K* of 40

(1) End of Year	(2) Exponential Growth	(3) Logistic Growth	Yearly Addition
1	4	4	4
2	8	8	6
3	16	14	9
4	32	23	10
5	64	33	6
6	128	39	1
7	256	40	0
8	512	40	0
9	1,024	40	0
10	2,048	40	0

Note: Column 3 is the yearly addition for logistic growth. Columns 2 and 3 are rounded to the nearest whole number.

the example, the inflection point occurs somewhere between Year 4 and Year 5. The sigmoid growth curve is a conceptual model of what could happen but contains many marginally questionable assumptions, such as

- Mortality is entirely dependent on population density.
- Natality and mortality are continuous, not discontinuous, functions.
- The environment never changes.
- All individuals in the population react in the same way to changes in density.
- The expression $(K - P)/K$ has a biological basis.

The causes underlying the change from a J-shaped to an S-shaped growth curve have led to several differing schools of thought about the effects of animal density on animal populations. One way to analyze the effects is to categorize the influences into density-dependent and density-independent factors.

Independent factors are those affecting the population regardless of size—the hazards, diseases, predators, and resources. *Density-dependent factors,* on the other hand, are those factors affecting the ability of individuals to reproduce, such as stress, behavior, and the shortage of resources as a result of competition. In the wild, the rate of increase of a population is reduced not by any single density-dependent or density-independent factor but by a combination of both. *Population regulation* describes a combination of feedback mechanisms that bring about resilience or equilibrium of a population. Density-independent factors do not regulate populations but can reduce numbers on a yearly basis. The important point is that populations of any species cannot increase indefinitely without some regulating factor or factors intervening to control the density. Chapman (1931) defined the intervening factors as environmental resistance.

With so many assumptions and difficulties in measuring population factors, it hardly seems worthwhile to view the inhibited growth model as a useful tool for wildlife management. Most wildlife managers will not have occasion to use the formulas or to determine an r value because virtually all of the populations with which they work are already somewhere between K/2 and the carrying capacity. Despite the theoretical difficulties, growth models provide wildlife managers an intellectual framework to roughly determine the limits within which management activities can be conducted and to understand new approaches as they are developed.

Intuitively, the initial growth in both the uninhibited and inhibited models makes sense because animals reproduce and add young to a population, but eventually something happens to restrict that growth. The growth models provide the foundation for concluding that if a given population was once high and is now low, then presented the right conditions that population can return to high densities rather rapidly. Wildlife management seeks to discover what that mix of right conditions is. One of the great practical benefits of understanding growth curves and their characteristics is that they give the wildlife manager a reference for understanding and discussing population problems.

2.6.9 CARRYING CAPACITY

Critical to the inhibited growth model is the concept of *carrying capacity* (K). The idea that an area can support only so many animals is well established in wildlife literature, but an exact meaning of the term has defied attempts at clarification (Edwards and Fowle 1955). Most managers would agree that carrying capacity is a balance between vegetation and animals characterized by an expression of animal density (Caughley 1979). While the idea of carrying capacity is simple, the actual definition is more complicated because some type of constraint is needed for quantification. Carrying capacity has been described and used in many ways in wildlife management, but primarily three ways have been used in the past (Dasmann 1981):

- The number of animals of a given species actually supported by a habitat, measured over a period of years
- The upper limit of population growth in a habitat above which no further increase can be sustained
- The number of animals a habitat can maintain in a healthy, vigorous condition

Most of the ideas about wildlife carrying capacity were first developed by range managers; the emphasis was on the density of animals. In the case of livestock, the term *grazing capacity* seems more appropriate (Stoddart et al. 1975). Carrying capacity is largely a concept rather than a real, quantifiable entity. Extending the concept to amphibians, birds, small mammals, and reptiles is not common, but if carrying capacity is real, then an extension to these life forms should be possible.

As a population produces young each year, animal numbers (as for deer and elk) may exceed the year-long carrying capacity until some environmental factor checks the population increase. As the habitat quality of a given vegetation stage increases for a particular species, the carrying capacity for that species increases. Carrying capacity is not a fixed factor of the land and will change as food, cover, and water change with the seasons or over long time periods, so the process is dynamic, not static as is often suggested by the many definitions. For herbivorous animals such as deer, carrying capacity is difficult to determine, and if capacity is exceeded for several years, food and cover can be affected adversely. The problem is that of detecting when animal density is starting to exceed carrying capacity. Most ungulate populations cannot be maintained at a level above carrying capacity for long periods without damage to the resource base.

In addition to reducing the impact of the factors that reduce the value of r in the inhibited growth model, the ecosystem manager's job is to keep animal numbers at a level that will not damage the basic soil and plant resources. In reality, animal populations fluctuate above and below carrying capacity because of the environmental factors of hazards, predators, disease, resources, and humans. A continued fluctuation suggests a process leading to a balance or equilibrium between resources (primarily vegetation) and animal density. This fluctuation of population numbers might be termed the process of *dynamic equilibrium* (the same term is used in plant and animal succession and the river continuum concept) in place of the term carrying capacity (Caughley 1979). There are three parts to the process of reaching dynamic equilibrium for herbivores:

- Growth response of plants: the rate of increase of plant biomass as a function of plant density
- Functional response of herbivores: the rate of intake per animal as a function of plant density
- Numerical response of herbivores: the rate of increase per animal as a function of plant density

When equilibrium is reached between plant growth and animal density in the absence of hunting, then the *ecological carrying capacity* has been reached. By interjecting hunting, a new state of equilibrium is achieved that is below the ecological capacity; this is termed *economic carrying capacity*. The economic carrying capacity is determined by quantifying the density of animals desired to meet certain objectives.

2.6.10 LIFE TABLE

So far, the discussion of population dynamics has centered on natality, but what about mortality? What tools are available to help understand the changes in a cohort of animals over their lifetime? Fortunately, the study of human populations (demography) has provided techniques to deal with mortality by specific age classes. These techniques were developed by life insurance companies, which were forced to estimate how long a human in a certain age class will live. Wildlife managers can use these same techniques to deal with questions of mortality among wildlife species.

A life table follows a cohort of animals through its various outcomes until the last member has died. A life table developed in this manner is an *age-specific* or *dynamic* life table. However, biologists seldom have the luxury of studying a group of animals all born at the same time. The next-best option is to reconstruct from either census data or death records what happens to groups of animals. These data lead to a

TABLE 2.9

**Composite Life Table for the Tassel-Eared Squirrel
(Subspecies Kaibabensis) (Combined Sexes)**

Age x	Frequency fx[a]	Survival lx[b]	Mortality dx	Mortality Rate qx	Survival Rate
0–1	58	1,000	552	0.552	0.448
1–2	26	448	207	0.462	0.538
2–3	14	241	69	0.286	0.714
3–4	10	172	86	0.500	0.500
4–5	5	86	52	0.605	0.395
5–6	2	34	34	1.000	

Source: My original data.

[a] Number of animals remaining in each age class from an initial population of 58.

[b] Number of animals converted to a base cohort of 1,000.

time-specific life table in which animals in one period of time are captured and their distribution of ages determined. The mortality factor is calculated from the distribution of the age classes based on the assumption that a stable age distribution exists.

A third kind of life table is a *composite* life table developed from mark-recapture data for two or more periods of time. Techniques for constructing this type of life table can be found in the *Wildlife Management Techniques Manual* (Schemnitz 1980). By way of example, Table 2.9 shows results for a cohort of 58 Kaibab squirrels trapped for a period of 8 yr, 2 of which lived to be 6 yr old (fx column) (my original data). The tassel-eared squirrel has a litter averaging about 3–5. While fecundity is high, so is the mortality in the 0–1 age class, which is 55%; decreases to mid-age (2–3); and then increases in the older (4+ years) age classes. This mortality rate indicates an unhunted population in which some adults are able to approach the limits of physiological longevity (about 8 yr for a tassel-eared squirrel). Caughley (1966) suggested that mortality in mammals follows a "U-shaped" trend with age; mortality in the Kaibab squirrel life table follows this pattern. Life tables, while a useful tool in analyzing a population, are difficult to construct, and the data are expensive to collect; therefore, they are almost never used in real management situations.

2.6.11 FLUCTUATIONS

Fluctuate in wildlife management means a change in quantity in a short time. The most common type of fluctuation is the yearly peaks brought about by high populations of small animals born in the spring to the winter trough when animals die due to various environmental factors (Figure 2.4). While there is yearly change in most small animal populations, for example, cottontails and quail, there is no great departure from the peaks and troughs between years; therefore, the population tends to oscillate over time, giving the curve of annual change a flat appearance. Leopold (1933) has termed this type of annual change a *stable population*.

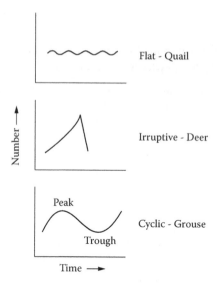

FIGURE 2.4 Three types of fluctuation.

The populations of some animal species fluctuate through peaks and troughs separated by several years in a predictable pattern; such fluctuations are known as *cycles*. Two distinct multispecies cycles have been described: a short cycle of 3–4 yr and a long cycle of 9–10 yr. Lemmings and voles in the Arctic and subarctic, along with their predators the Arctic fox, red fox, and snowy owl, are short-cycle species. Elton (1942) was the first to discover and report on the short cycle. Short cycles are exhibited by animals having a limited breeding season or short life cycle and occur in simple ecosystems with communities that include only a few species.

The long cycle became evident from the Hudson Bay Company fur records, which indicated a cycle of 9–10 yr for the Canada lynx (Elton and Nicholson 1942). A snowshoe hare population increase was accompanied by an increase in the lynx population, which lagged the snowshoe peaks and valleys. Other long-cycle species include ruffed grouse, muskrat, bobwhite, and pheasants (Errington 1945; Hewett 1954; Kozicky et al. 1955).

A second type of fluctuation is an *irruption* (or eruption), which is marked by a sharp increase in a short time followed by a population crash with a return to stability. The Kaibab deer herd crash of 1940 is considered the classic example of an irruption (Leopold 1943; Leopold et al. 1947), but irruptions have been reported for jackrabbits (Wagner and Stoddart 1972), quail (Gullion 1960), and thar in New Zealand (Caughley 1970). Irruptions can occur when new habitat is created by disturbance, as happened in California in the Devil's Garden deer herd (Salwasser 1979).

Theoretically, fluctuations can be caused by any of the external factors. In addition to these factors, however, certain other events or conditions seem to correlate with high populations and have been hypothesized, at least in part, to be the cause of cycles. Some of these theories are the ozone theory, sunspot theory, stress theory, predator–prey theory, and random theory (Christian 1950; Calhoun 1952; Cole 1954;

Krebs et al. 1973). At one time, cycles were a subject of major concern and study, but the topic now seems to be of little interest. This is because most biologists accept the validity of crude cycles but also know that cycles are not a factor that can be managed. However, this is not the case for irruptive species such as deer, for which increases can be detected and management action taken before a crash occurs.

2.6.12 Demographic Vigor

Conceptual Equation 1.1 identified six factors that act directly on the ability of an animal or a population to survive and reproduce. It is clear that the value r in the exponent of Population Equation 2.5 has a tremendous effect on a population because it increases animal numbers exponentially, as in compound interest. What is it, then, that affects r? The rate of increase r integrates the six factors of the equation into a single value characterizing the probabilities of a population to survive and reproduce. It reflects the action of environmental influences and summarizes quantitatively the vigor or "health" of a population in given circumstances, and it is a quantification of biotic potential.

Because of the difficulties inherent in establishing a real value for r, biologists must search for a reasonable indicator to use in management decisions. For years, wildlife biologists in the United States and elsewhere have been counting cow:calf ratios, doe:fawn ratios, number of eggs hatched, hen:poult ratios, and other measures of productivity. These types of ratios and counts, if done consistently in the same environmental conditions, can be a good indicator of r as demographic vigor and can form the basis for practical management decisions. However, it is possible that populations with a high r value can still decrease because female young may not live long enough to reproduce and add offspring to the population. If reproduction is not a problem, then the other factors of disease, hazards, predators, and so on, may be the ones keeping a population in a declining state.

2.7 RESOURCES

Food, cover, water, and space habitat components are necessary to maintain basic physiological functions of an animal to survive and reproduce and subsequently for the offspring to become members of a population. Because of this important fact, managers spend most of their time dealing with these requirements to maintain stable wildlife populations. The need for food, cover, water, and space for a particular species is as variable as the size, mobility, and niche of the species. Successful managers understand this variability and so direct their management activities to provide diverse resources to meet the needs of a variety of species. In 1798, Malthus proposed the *principle of population,* in which he hypothesized that all animal species tend to reproduce to a point in excess of available food resources and this will limit population density.

While it is possible to control populations by reducing cover, food is more likely to be a limiting factor because it maintains body heat. The critical components necessary for an animal to survive are found in the resource factor of the centrum. Amounts of food and cover are easy to quantify in a stand using the common inventory techniques to measure trees, shrubs, herbaceous material, and coarse wood debris. However, it is not so easy to quantify or determine which forest stand components are used by which species and how important these components are in terms of a quality measurement such as poor to excellent.

2.7.1 FOOD

The food resource involves not only the availability of food necessary to obtain energy but also the conservation of this energy once it has been obtained. The consumption of food is determined by availability and palatability. *Availability* refers to what is obtainable during the various seasons of the year; for example, Turkowski (1980) found that in ponderosa pine in Arizona the diet of a coyote will vary from a use of juniper berries, deer, and elk in the winter to small mammals and arthropods in the summer, only to return to juniper berries and deer in the fall (Table 2.10). Even though the coyote is a carnivore, plant material consisting primarily of juniper berries is an important food item and is used throughout the year.

Availability also relates to the physical location of a food item, for example, browse out of reach of deer and elk because of a physical barrier or vertical location of leaves and twigs that are out of reach. In many cases, as for large animals, the size of the food item is too small to equal the energy it takes to secure the item. Food palatability includes qualities of preference, nutritive content, and digestibility.

TABLE 2.10
Seasonal Food Habits of Coyotes in Ponderosa Pine Forests in Arizona

Season	Item[a]
Winter	Juniper berries
	Deer and elk
	Small mammals
Spring	Juniper berries
	Small mammals
	Deer and elk
	Insects
Summer	Small mammals
	Arthropods
	Insects
	Juniper berries
Fall	Juniper berries
	Deer and elk
	Arthropods

[a] Listed in order of use by season.

Preference describes the hierarchical ordering of food, the highest being the item selected more frequently than any other. This hierarchical priority can generally be ordered as

- High value (preferred): eaten most frequently and in large amounts when available from an array of food plants
- Moderate value (staple): second choice but still provides nutrition in moderate amounts when an array of food plants is available
- Low value (stuffing-emergency): fulfills short-term needs but generally low in nutritional value; generally eaten in small amounts unless it is the only food plant available

Food habits of various animals indicate that a variety of plants is consumed, and even when high-value plants are available and being eaten, small amounts of low-value plants remain in the diet. Nutrients are needed by all living organisms for body maintenance and reproduction, but requirements vary by species. More information is available on the food requirements of wildlife species than any other topic in the wildlife profession. One needs only to read the *Journal of Wildlife Management* from the early 1940s through the 1960s to obtain an appreciation of the effort expended by research biologists and graduate students in determining food requirements of wildlife species.

Comprehensive publications are now available describing food habits at the broad (Martin et al. 1951), regional (Patton and Ertle 1982), and local (Stephenson 1974) levels. The nutritional requirements of game animals have been most extensively studied because the wildlife profession generally began by managing for hunted species such as deer, elk, turkey, rabbits, squirrels, and quail. Due to the great variation in the nutritive content of plants, wildlife biologists are still seeking a single index of plant nutritional value.

Animals need vitamins, fats, proteins, minerals, and carbohydrates. But, protein seems to be the most likely indicator of plant nutritional quality. Some of the factors affecting the crude protein content of plants are soil moisture, soil nutrients, canopy closure, grazing intensity, and burning. While protein is a good indicator of plant quality, the complete nutritive value of a plant must be compared with the requirements of a species. Table 2.11 presents examples of the daily protein requirements for a few animals for comparison.

TABLE 2.11
Protein Requirements of Wild Animals

Species	Protein (%)	Reference
White-tailed deer	6.7–16	French et al. 1956
Wild turkey	15–20	Halls 1970
Bobwhite quail	12–28	Nestler 1949

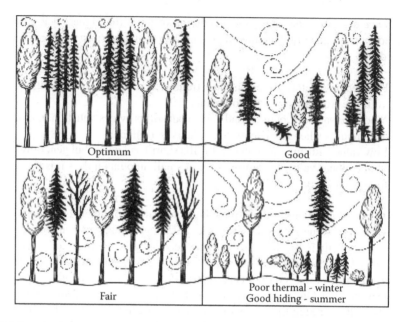

FIGURE 2.5 Vegetation structures provide different qualities of cover for white-tailed deer. (From Verme, L. J., *J. Forest.* 63:523–529, 1965.)

2.7.2 COVER

Once energy has been obtained through a proper diet, an animal must then conserve this energy through some form of cover. *Cover* is a term that in natural resource management refers to a layer of vegetation over a substrate. In forestry, there is forest cover type; in range management, there is ground cover, but in wildlife management cover includes more than vegetation. *Wildlife cover* is a place of shelter or conceal-ment, which can include vegetation or any type of physical structure, such as rocks, logs, stumps, and snags. Cover can be both shelter from weather and predators and concealment for resting, calving, or thermal regulation. The cover requirements of individual species are as variable as food requirements for species.

Cover is furnished by vegetative structure (trees, shrubs, grasses, and forbs) as well as by plant stages within life forms (seedling, sapling, etc.), topographic features (aspect, hills, valleys, soil, etc.), and water (Figure 2.5). Although cover provides protection, there are special types of cover associated with specific functions for each wildlife species. Some of the cover types that have been identified are thermal (heat and cold) cover for deer and elk, roosting cover for turkeys, nesting cover for grouse, resting cover for quail, and escape cover for squirrels. Thermal cover for elk has been a recommendation by biologists for many years because of the moderat-ing effects of dense forest canopies. However, during a 4-yr test of thermal cover in Oregon, Cook et al. (1998) found that thermal cover did not appreciably enhance the energetic benefits of elk and was not a suitable solution for inadequate forage. The

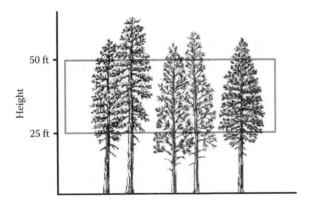

FIGURE 2.6 Nest zone of the tassel-eared squirrel in Arizona ponderosa pine.

authors recommended that elk biologists should refocus their attention to the influences of forest management on forage resources and related production potential of forest successional stages.

Like food, cover is subject to annual and long-term changes (ecological time), as in the change in location of forest stands. Care must be taken in describing cover in general terms; for example, thermal cover for elk in the summer may not exist in the same stand in the winter if the stand is composed of deciduous trees. In addition to general types of cover, for some species there is a subset that is used for survival purposes, for example, calving or for rearing young in a nest.

Tassel-eared squirrels build a nest in a single ponderosa pine, where young are born. This nest occurs in a tree surrounded by a group of trees. The tree group and individual nest tree provide cover from weather and predators, but the nest provides protection for the survival of newborn squirrels (Figure 2.6). The nest is a microsite within the cover area.

Not all species have specific cover requirements, but some types of cover that are used are rock crevices, tree cavities, hollow logs, and ground dens. When discussing cover and requirements of specific species, *caution* is urged in *not* transferring the requirements across geographical boundaries or vegetation types unless studies or experience have shown that the requirements are similar in the hierarchical levels being compared.

2.7.3 WATER

Water is a resource necessary for animals to survive and reproduce, and like nutritional requirements, water requirements vary by species (Table 2.12). Water is needed for metabolism (drinking) by different species of terrestrial mammals, as a medium for waterfowl reproduction and rearing of young, and as cover for fish and amphibians such as frogs. Species needing free water for metabolism must find it in their home range. Examples of research to determine water requirements of wild species are provided for antelope (Wesley et al. 1970) and for bighorn sheep (Turner 1973). Species can adapt to regions with low precipitation by conserving water, such

TABLE 2.12
Water Requirement of Some Common Animal Species

Species	Liters/Day	(Gallons)/Day	Spacing, km (mi)	Reference
Range cattle	28–38	(7.4–10)		Stoddart et al. 1975
Sheep	0.95–5.7	(0.2–1.5)		Stoddart et al. 1975
Mule deer	2.3	(0.6)	1.6–4.8 (1–3)	Bissell et al. 1955
Antelope	3.8–7.6	(1–2)	3.2–4.8 (2–3)	Payne and Copes 1986
Elk	18.9–30	(5–8)	1.6–4.8 (1–3)	Payne and Copes 1986
Quail, cov	7.7	(2.05)	0.8–1.6 (.5–1)	Payne and Copes 1986
Turkey, flo	21.02	(5.55)	0.6–3.2 (1–2)	Payne and Copes 1986
Songbirds, gro	3.8–7.6	(1–2)	0.4–0.8 (0.25–0.5)	Payne and Copes 1986

Note: cov, covey; flo, flock; gro, group.

as rodents going underground or deer and elk seeking shade to avoid heat; by morphological adaptations (body size and shape as in the jackrabbit); or by using metabolic water (kangaroo rat).

Research has shown that precipitation in the form of rainfall has an effect on the distribution and reproduction of some species; for example, in Arizona the number of young Gambel's quail is related to precipitation (Figure 2.7); in South Carolina, drought reduced the number of bobwhite young per female (Rosene 1969); in California, when rainfall is less than 6.3 in, the number of young California quail are reduced (McMillan 1964); and deer densities in Texas are related to the precipitation of the previous year (Teer et al. 1965). Many factors affect the distribution, availability, and potability of water for use by wildlife and, excluding rainfall patterns, most result from humankind's activities, such as dams and reservoirs, livestock use, wetland drainage, and pollution in many

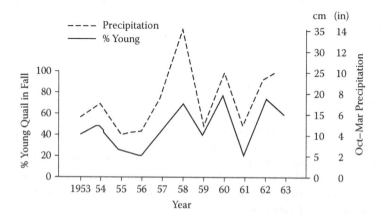

FIGURE 2.7 Number of young quail is related to precipitation. (From Gallizioli, S., *Quail research in Arizona,* Arizona Game and Fish Department, Phoenix, 1965.)

forms (sewage, acid rain, chemicals, oil spills, etc.). All these are part of the web of indirect factors.

In studies to determine the food and water requirements of a species, investigators must consider the following questions:

- What is available by season?
- What is accessible (height of browse plants, plants in rocky areas, water holes, lakes, etc.)?
- What is palatable and potable?
- What nutritional levels of plants and amounts of water are needed for physiological processes in each season?
- What is the individual adapted to eating (physical adaptations)?

2.7.4 SPACE

Space is a forest component that sometimes gets lost in the planning, assessment, and management process, particularly as it relates to space used by forest wildlife. All wildlife species live in an area that defines their habitats. Each area contains varying densities of the species, but in some areas, patches are unoccupied. Occupied and unoccupied areas across the continent make up the *geographical range*. A geographical range may be large, as in the case for white-tailed deer (45 states), or small, such as that of the Jemez Mountain salamander (one county). The geographical range provides an indication of the adaptability of a species to various climatic, topographic, and vegetative factors across latitudes and longitudes.

Within the geographical range of a given species, local and regional movements occur that coincide with the onset of fall and spring. This two-way movement is *migration*. Migrations are in some cases vertical, as in elk migrating up mountains in summer to avoid heat or to obtain high-quality food and thereafter descending to lower elevations in winter to snow-free areas where food is again available. Mule deer and moose also follow vertical movement patterns in many parts of North America.

Probably the best-known migrations are flights of waterfowl (Linduska 1964). These yearly movements of waterfowl between breeding grounds in northern Canada and their winter grounds largely in the United States and Mexico have resulted in the delineation of regional management areas based on seven migratory flyways: Atlantic, Mississippi, central, Pacific, Alaska, Canada, and Mexico. The Canadian and U.S. Fish and Wildlife Services have established a system of waterfowl refuges to provide food and cover for migrating birds. Migrations occur not only in terrestrial species but also in fish. The migration of salmon from freshwater to the sea and again to freshwater streams to spawn is well known.

2.7.5 MOBILITY

Mobility of an individual defines a *home range* (Burt 1943). Home range delineates the activities of an animal within a boundary that can be any shape extending from approximately circular to elliptical. The size of area used by mammals correlates with body weight and is roughly related to reproduction characteristics (Emlen

TABLE 2.13

Averages for Home Range of Mammals

Species	Hectares	Acres	Source
Ground squirrels	1	2	E. Yensen and P. W. Sherman
Cottontail	2	5	J. A. Chapman and J. A. Litvaitis
Raccoon	150	370	S. D. Gehrt
Gray fox	300	741	B. L. Cypher
White-tailed deer	350	865	K. V. Miller et al.
Red fox	1,260	3,112	B. L. Cypher
Coyote	1,400	3,458	M. Berkoff and E. M. Gese
Mule deer	2,500	6,175	R. J. Mackie et al.
Black bear	2,600	6,422	M. R. Melton
Elk	2,700	6,669	J. M. Peek
Moose	5,000	13,350	R. T. Bowyer
Bobcat	13,200	32,604	E. M. Anderson and M. J. Lovallo
Brown bear	29,300	73,371	C. C. Schwartz et al.
Cougar	30,000	74,100	B. M. Pierce and V. C. Bleich

Note: Sources are in Feldhamer et al. 2003.

1973). Species of low mobility are more at risk from habitat destruction than highly mobile species because of the scale of their space needs in comparison to management activities, such as timber harvesting.

As a result of advancements in radio telemetry, there is information available to define home range for a number of species (Table 2.13). One part of the home range is the daily *cruising radius* or the distance an animal travels in 1 day in search of food and cover. This daily distance over a period of time may define a *center of activity*. For example, a tassel-eared squirrel with a radio telemetry collar had a home range of about 110 ac but spent the most time in a 17-ac area. The center of activity is generally an area where the habitat is higher in quality, but not every animal has a center of activity.

In recent years, attention has been given to the size of an area that an animal uses, and this has developed into a generalization of relative distinction between two animal groups. Species with low mobility, high fecundity, small size, and all their requirements met in identifiable vegetation communities are *habitat specialists* that have narrow ecological amplitudes (also called *alpha species*) (Harris 1988). Species with large size, a low fecundity rate, high mobility, and a large landscape requirement are *habitat generalists* that have wide ecological amplitudes (also called *gamma species*). Between these two groups are animals that are *intermediate (beta species)* in their characteristics and habitat use. Figure 2.8 depicts a visual representation of the reproductive characteristics of an animal with size and mobility.

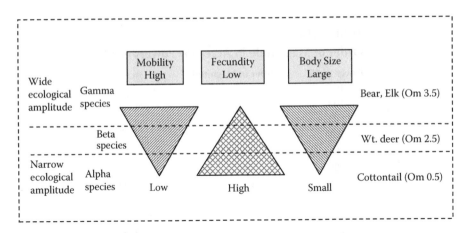

FIGURE 2.8 Reproductive characteristics associated with mobility and body size.

2.7.6 ORDER OF MAGNITUDE

Order of magnitude (Om) is a technique that can be used to compare two numbers of different categories on roughly the same scale. It is the number of powers of 10 contained in the number. Om is an approximate position on a logarithmic scale. An Om of a variable with an exact value that is unknown (such as home range) is an estimate rounded to the nearest power of 10.

For forestry, the unit of measure is size of

- Management units,
- Stands,
- Cut areas, and so on.

The unit of measure for wildlife is

- Home range and
- The management unit size needed to maintain viable populations of different classes of wildlife: endangered and threatened species, nongame and migratory birds, or game animals.

Two numbers with the same Om have the same scale; the scale contains values of a fixed ratio to the preceding class. The ratio most commonly used is 10. For example, an Om of 3 is 10^3 or 1,000 ac. An Om of 4 is 10^4 or 10,000 ac (Table 2.14). Based on the need to consider home ranges of many species, the need to preserve areas for interior species, and the desires of society for areas with old forest conditions, it is realistic that Om greater than 4.5 (50,000 to 100,000 ac) might be appropriate area sizes for landscape-level planning.

Figure 2.9 shows a comparison of home range to area based on Om as the common scale; an Om 4 includes the areas from 4 to 4. 5. A landscape of Om 4.5

TABLE 2.14

Order of Magnitude in Acres for Powers of 10

Power of 10	Order of Magnitude	Acres
10^{-2}	−2	0.01
10^{-1}	−1	0.1
10^{0}	0	1
$10^{0.5}$	0.5	3
10^{1}	1	10
$10^{1.5}$	1.5	32
10^{2}	2	100
$10^{2.5}$	2.5	316
10^{3}	3	1,000
$10^{3.5}$	3.5	3,162
10^{4}	4	10,000
$10^{4.5}$	4.5	31,623
10^{5}	5	100,000
10^{6}	6	1,000,000

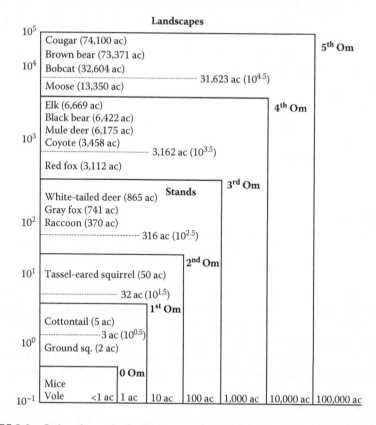

FIGURE 2.9 Order of magnitude (Om) comparison of animal home range to size of area.

would include areas from 4.5 to 5. Home ranges of individuals overlap, and this has to be considered when determining the size of a management area needed to maintain a viable population. The Om only considers space and not actual habitat within that space. The interpretation is that the space in a home range of Om 4 will encompass many species, with ranges from less than 1 to 99,999 ac. One advantage in using Om is that because home ranges are highly variable for a species, the acreages for home range estimates are only approximate but do not have to be exact in an Om category.

KNOWLEDGE ENHANCEMENT READING

Bolen, E. G., and W. L. Robinson. 2003. *Wildlife ecology and management.* Prentice-Hall, Pearson Education, Upper Saddle River, NJ.

Bookhout, T. A. (Ed.). 1996. *Research and management techniques for wildlife and habitats.* Wildlife Society, Bethesda, MD.

Caughley, G., and A. R. E. Sinclair. 1994. *Wildlife biology and management.* Blackwell Scientific, Boston.

Krebs, C. J. 1999. *Ecological methods.* Addison-Wesley Longman, Menlo Park, CA.

McComb, B. C. 2008. *Wildlife habitat management: Concepts and applications in forestry.* CRC Press, Taylor and Francis Group, Boca Raton, FL.

Payne, N. F., and F. C. Bryant. 1994. *Techniques for wildlife habitat management of uplands.* Biological Resource Management Series. McGraw-Hill, New York.

Williams, B. K., J. D. Nichols, and M. J. Conroy. 2001. *Analysis and management of animal populations.* Academic Press, San Diego, CA.

REFERENCES

Allen, D. L. 1974. *Our wildlife legacy.* Funk and Wagnalls, New York.

Allen, D. L. 1979. *Wolves of Minong: Their vital role in a wild community.* Houghton Mifflin, Boston.

Andrewartha, H. G., and L. C. Birch. 1984. *The ecological web: More on the distribution and abundance of animals.* University of Chicago Press.

Arizona Department of Health Services. 2007. September news release on plague in Arizona. State of Arizona, Phoenix.

Baker, W. W. 1974. Longevity of lightning struck trees and notes on wildlife use. *Tall Timbers Fire Ecol. Conf.* 13:497–504.

Bendell, J. F. 1974. Effects of fire on birds and mammals. In T. T. Kozlowski and C. E. Ahlgren (Eds.), *Fire and ecosystems,* pp. 73–138. Academic Press, New York.

Bissell, H. D., B. Harris, H. Strong, et al. 1955. The digestibility of certain natural and artificial foods eaten by deer in California. *Calif. Fish Game* 41:57–78.

Bock, C. E., and J. F. Lynch. 1970. Breeding bird populations of burned and unburned conifer forest in the Sierra Nevada. *Condor* 72:182–189.

Bowyer, R. T., M. E. Shea, and S. A. McKenna. 1986. The role of winter severity and population density in regulating northern populations of deer. In J.A. Bissonette (Ed.). *Is good forestry good wildlife management?* pp. 193–204. Maine Agriculture Experimental Station, Miscellaneous Publication Number 689. Orono, ME.

Brown, J. L. 1969. Territorial behavior and population regulation in birds: A review and re-evaluation. *Wilson Bull.* 81:293–329.

Buech, R. R., K. Siderits, R. E. Radtke, et al. 1977. *Small mammal populations after a wildfire in northeast Minnesota.* Research Paper NC-151. U.S. Forest Service, St. Paul, MN.

Buechner, H. K. 1950. Life history, ecology, and range use of the pronghorn antelope in Trans-Pecos Texas. *Am. Midl. Nat.* 43:257–354.

Burt, W. H. 1943. Territoriality and home range concepts as applied to mammals. *J. Mammal.* 24:346–352.

Calhoun, J. B. 1952. The social aspects of population dynamics. *J. Mammal.* 33:139–159.

Carson, R. 1962. *Silent spring.* Fawcett, Greenwich, CT.

Caughley, G. 1966. Mortality patterns in mammals. *Ecology* 47:906–918.

Caughley, G. 1970. Eruption of ungulate populations with emphasis on Himalayan Thar in New Zealand. *Ecology* 51:53–72.

Caughley, G. 1977. *Analysis of vertebrate populations.* John Wiley and Sons, New York.

Caughley, G. 1979. What is this thing called carrying capacity? In M. S. Boyce and L. D. Hayden-Wing (Eds.), *North American elk: Ecology behavior, and management,* pp. 2–8. University of Wyoming, Laramie.

Chapman, R. N. 1931. *Animal ecology.* McGraw-Hill, New York.

Chew, R. M., B. B. Butterworth, and R. Grechman. 1958. The effects of fire on the small mammal populations of chaparral. *J. Mammal.* 40:253.

Christian, J. J. 1950. The adreno-pituitary system and population cycles in mammals. *J. Mammal.* 31:246–260.

Cole, L. C. 1954. Some features of random population cycles. *J. Wildl. Manage.* 18:2–24.

Cook, J. G., L. L. Irwin, L. D. Bryant, et al. 1998. *Relations of forest cover and condition of elk: A test of the thermal cover hypothesis in summer and winter.* Wildlife Monograph Number 41. Wildlife Society, Washington, DC.

Cook, S. F. 1959. The effects of fire on a population of small rodents. *Ecology* 40:102–108.

Council on Environmental Quality. 1981. *Environmental trends.* Executive Office of the President, Washington, DC.

Dasmann, R. F. 1981. *Wildlife biology.* John Wiley and Sons, New York.

Daubenmire, R. 1968. Ecology of fire in grasslands. In *Advances in ecological research,* Vol. 5, pp. 209–273. Academic Press, New York.

Davidson, W. R., and V. F. Nettles. 1988. *Field manual of wildlife diseases in the southeastern United States.* Southeastern Cooperative Disease Study, University of Georgia, Athens.

Davies, N. B. 1978. Ecological questions about territorial behavior. In J. R. Krebs and N. B. Davies (Eds.), *Behavioral ecology: An evolutionary approach,* pp. 317–335. Sinauer Associates, Sunderland, MA.

Decker, D. J., T. L. Brown, and W. Sarbello. 1981. Attitudes of residents in the peripheral Adirondacks toward illegally killing deer. *N.Y. Fish Game J.* 28:73–80.

Dodd, N. L., J. W. Gagnon, S. Boe, et al. 2006. Characteristics of elk-vehicle collisions and comparison to GPS-determined highway crossing patterns. In C. L. Irwin, P. Garrett, and K. P. McDermott (Eds.), *Proceedings of the 2005 international conference on ecology and transportation,* pp. 461–477. Center for Transportation and the Environment, North Carolina State University, Raleigh.

Edwards, R. Y. 1954. Fire and the decline of a mountain caribou herd. *J. Wildl. Manage.* 18:521–526.

Edwards, R. Y., and C. D. Fowle. 1955. The concept of carrying capacity. *Trans. N. Am. Wildl. Conf.* 20:589–602.

Einarsen, A. S. 1945. Some factors affecting ring-neck pheasant population density. *Murrelet* 26:2–9, 39–44.

Elton, C. S. 1942. *Voles, mice and lemmings.* Clarendon Press, London.

Elton, C. S., and M. Nicholson. 1942. The ten-year cycle in numbers of lynx. *Can. J. Animal Ecol.* 11:215–244.

Emlen, J. M. 1973. *Ecology: An evolutionary approach.* Addison-Wesley, Reading, MA.

Errington, P. L. 1945. Some contributions of a 15-year study of the northern bobwhite to a knowledge of a population phenomena. *Ecol. Monogr.* 15:1–34.

Errington, P. L. 1967. *Of predation and life.* Iowa State University Press, Ames.

Feldhamer, G. A., B. C. Thompson, and J. A. Chapman (Eds.). 2003. *Wild mammals of North America, biology, management, and conservation.* Johns Hopkins University Press, Baltimore, MD.

Fox, M. W. 1980. *The soul of the wolf.* Little, Brown, Boston.

French, C. E., L. C. McEwen, N. D. MacGruder, et al. 1956. Nutrient requirements for growth and antler development in the white-tailed deer. *J. Wildl. Manage.* 20:221–232.

Gallizioli, S. 1965. *Quail research in Arizona.* Arizona Game and Fish Department, Phoenix.

Gilbert, F. E., and D. G. Dodds. 2001. *The philosophy and practice of wildlife management.* Krieger, Malabar, FL.

Grinnell, J. 1943. *Philosophy of nature.* University of California Press, Berkeley.

Gullion, G. W. 1960. The ecology of Gambel's quail in Nevada and the arid Southwest. *Ecology* 14:518–536.

Hailey, T. L., J. W. Thomas, and R. M. Robinson, 1966. Pronghorn die-off in Trans-Pecos Texas. *J. Wildl. Manage.* 30:488–496.

Hakala, J. B., R. K. Seemel, R. A. Richey, et al. 1971. Fires effects and rehabilitation methods: Swanson-Russian River fires. In C. W. Slaughter, R. J. Burns, and G. M. Hansen (Eds.), *Proceedings: A symposium on fire in the northern environment,* pp. 87–89. U.S. Forest Service, Portland, OR.

Halls, L. K. 1970. Nutrient requirements of livestock and game. In H. A. Paulsen Jr., and E. H. Reid (Eds.), *Range and wildlife habitat evaluation,* pp. 10–24. Miscellaneous Publication Number 1147. U.S. Forest Service, Washington, DC.

Harris, L. D. 1988. Reconsideration of the habitat concept. *Trans. N. Am. Wildl. Nat. Resources Conf.* 53:137–144.

Hewett, O. (Ed.). 1954. A symposium on cycles in animal populations. *J. Wildl. Manage.* 18:1–112.

Hooper, R. G., J. Watson, and E. F. Escano. 1990. Hurricane Hugo's initial effects on red cockaded woodpeckers in the Francis Marion National Forest. In *Transactions, 55th North American wildlife and natural resources conference,* pp. 220–224. Wildlife Management Institute, Washington, DC.

Hornocker, M. G. 1969. Winter territoriality in mountain lions. *J. Wildl. Manage.* 33:457–464.

Howard, H. E. 1920. *Territory in bird life.* Dutton, New York.

Howard, W. E., R. L. Fenner, and H. E. Childs, Jr. 1959. Wildlife survival in brush burns. *J. Range Manage.* 12:230–234.

Karstad, L. 1979. Diseases of wildlife. In R. D. Teague and E. Decker (Eds.), *Wildlife conservation,* pp. 123–127. Wildlife Society, Washington, DC.

Kendeigh, S. C. 1941. Territorial and mating behavior of the house wren. *Illinois Biol. Monogr.* 10:1-20.

Klein, D. R. 1968. The introduction, increase, and crash of reindeer on St. Matthew Island. *J. Wildl. Manage.* 32:350–367.

Komarek, E. V., Sr. 1969. Fire and animal behavior. *Tall Timbers Fire Ecol. Conf.* 9:161–207.

Koya, C. M. 1977. Reproductive biology of rainbow and brown trout in a geothermally heated stream: The Firehole River of Yellowstone National Park. *Trans. Am. Fisheries Soc.* 106:354–361.

Kozicky, E. L., G. O. Hendrickson, and P. G. Homer. 1955. Weather and fall pheasant populations. *J. Wildl. Manage.* 19:136–142.

Kramp, B. A., D. R. Patton, and W. W. Brady. 1983. *The effects of fire on wildlife habitat and species.* Run Wild: Wildlife habitat relationships, SW Region Technical Report. U.S. Forest Service, Albuquerque, NM.

Krebs, C. J., M. S. Gaines, B. L. Keler, et al. 1973. Population cycles in small rodents. *Science* 179:35–41.

Lawrence, G. E. 1966. Ecology of vertebrate animals in relation to chaparral fire in the Sierra Nevada foothills. *Ecology* 47:278–291.

Leopold, A. 1933. *Game management.* Charles Scribner's Sons, New York.

Leopold, A. 1943. Wildlife in American culture. *J. Wildl. Manage.* 7:1–6.

Leopold, A., L. K. Sowls, and D. L. Spencer. 1947. A survey of over-populated deer ranges in the United States. *J. Wildl. Manage.* 11:162–177.

Leopold, A. S., S. A. Cain, C. M. Cottam, et al. 1964. Predator and rodent control in the United States. *Trans. N. Am. Wildl. Nat. Resources Conf.* 29:27–49.

Linduska, J. P. (Ed.). 1964. *Waterfowl tomorrow.* U.S. Fish and Wildlife Service, Washington, DC.

Lotka, A. J. 1925. *Elements of physical biology.* Williams and Wilkins, Baltimore, MD.

Lyon, L. J. 1984. Road effects and impacts on wildlife and fisheries. Paper presented at symposium on forest transportation, December 11–13, Casper, WY. U.S. Forest Service, Region 2, Denver, CO.

Lyon, L. J., H. S. Crawford, E. Czuhai, et al. 1978. *Effects of fire on fauna: A state-of-knowledge review.* General Technical Report 6. U.S. Forest Service, Washington, DC.

MacArthur, R. H., and E. O. Wilson. 1967. *The theory of island biogeography.* Princeton University Press, Princeton, NJ.

Malthus, T. R. 1798. *An essay on the principle of population.* Johnson, London.

Marcot, B. G. 1986. Concepts of risk analysis as applied to viable population assessment and planning. In B. A. Wilcox, P. F. Brussard, and B. G. Marcot (Eds.), *The management of viability: Theory, applications, and case studies*, pp. 89–101. Center for Conservation Biology, Department of Biological Science, Stanford University, Stanford, CA.

Martin, A. C., H. S. Zim, and A. L. Nelson. 1951. *American wildlife and plants: A guide to wildlife food habits.* Dover, New York.

McCulloch, C. Y., and R. L. Brown. 1986. *Rates and causes of mortality among radio collared mule deer of the Kaibab Plateau, 1978–1983.* Federal Aid in Wildlife Restoration Project W-78-R. Arizona Game and Fish Department, Phoenix.

McCullough, D. R. 1979. *The George Reserve deer herd: Population ecology of a K-selected species.* University of Michigan Press, Ann Arbor.

McMillan, I. I. 1964. Annual population changes in California quail. *J. Wildl. Manage.* 28:702–711.

McWhirter, N. 1981. *Guiness book of world records.* Sterling, New York.

Mech, L. D., L. Frenzel Jr., R. Meam, and J. Winship. 1971. *Ecological studies of the timber wolf in northeastern Minnesota.* Research Paper NC-52. U.S. Forest Service, St. Paul, MN.

Merriam-Webster. 2008. *Merriam-Webster's collegiate dictionary.* Merriam-Webster, Springfield, MA.

National Academy of Science. 1982. *Impacts of emerging agricultural trends on fish and wildlife habitat.* National Academy Press, Washington, DC.

Nestler, R. B. 1949. Nutrition of bobwhite quail. *J. Wildl. Manage.* 13:342–358.

Nice, M. 1941. The role of territory in bird life. *Am. Mid. Nat.* 26:441–487.

Oxley, D. J., M. B. Fennton, and G. R. Carmody 1974. The effects of roads on populations of small mammals. *J. Appl. Ecol.* 11:51–59.

Pattee, O. H., and S. K. Hennes. 1983. Bald eagles and waterfowl, the lead shot connection. *Trans. N. Am. Wildl. Nat. Resources Conf.* 48:230–237.

Patton, D. R., and M. Ertle. 1982. *Wildlife food plants of the Southwest.* Wildlife Unit Technical Report. U.S. Forest Service, SW Region, Albuquerque, NM.

Payne, N. F., and F. Copes. 1986. *Wildlife and fisheries habitat improvement handbook.* Wildlife and Fisheries Administrative Report. U.S. Forest Service, Washington, DC.

Perry, C., and R. Overly. 1976. Impact of roads on big game distribution in portions of the Blue Mountains of Washington. In J. M. Peek (Chair), *Proceedings: A symposium on elk, logging, and roads*, pp. 62–68. University of Idaho, Moscow.

Primack, R. B. 2006. *Essential of conservation biology.* Sinauer Associates, Sunderland, MA.

Rasmussen, D. I. 1941. Biotic communities of the Kaibab plateau. *Ecol. Monogr.* 3:229–275.

Rippe, D. J., and R. L. Rayburn. 1981. *Land use and big game population trends in Wyoming.* W/CRAM-81-W22. U.S. Fish and Wildlife Service, Washington, DC.

Roadkill statistics. 2005. *High Country News* [Paonia, CO]. Sidebar article, February 7.

Rosene, W. 1969. *The bobwhite quail: Its life and management.* Rutgers University Press, New Brunswick, NJ.

Salwasser, H. 1979. The ecology and management of the Devil's Garden interstate deer herd and its range. Ph.D. dissertation, University of California, Berkeley.

Schemnitz, S. D. 1980. *Wildlife management techniques manual.* Wildlife Society, Washington, DC.

Schoener, T. W. 1968. Sizes of feeding territories among birds. *Ecology* 49:123–141.

Schonewald-Cox, C. M. 1983. Guidelines to management: A beginning attempt. In C. M. Schonewald-Cox, S. M. Chambers, B. MacBryde, et al. (Eds.), *Genetics and conservation, a reference for managing wild animal and plant populations*, pp. 414–445. Benjamin-Cummings, Menlo Park, CA.

Severinghaus, C. W. 1972. Weather and deer populations. *Conservationist* 27:28–31.

Smith, T. M., and R. L. Smith. 2009. *Elements of ecology.* Benjamin Cummings, San Francisco.

Spight, T. M. 1967. Species diversity: A comment on the role of the predator. *Am. Nat.* 101:467–474.

Stephenson, R. L. 1974. Reproduction biology and food habits of Abert's squirrels in central Arizona. M.S. thesis, Arizona State University, Tempe.

Stoddart, L. A., A. D. Smith, and T. W. Box. 1975. *Range management.* McGraw-Hill, New York.

Swanston, D. N. 1980. *Influence of forest and rangeland management on anadromous fish habitat in western North America: Impacts of natural events.* General Technical Report PNW-104. U.S. Forest Service, Portland, OR.

Teer, J. G., J. W. Thomas, and E. A. Walker. 1965. *Ecology and management of white-tailed deer in the Llano Basin of Texas.* Wildlife Monograph Number 15. Wildlife Society, Washington, DC.

Trombulak, S. C., and C. A. Frissell. 2000. Review of ecological effects of roads on terrestrial and aquatic communities. *Cons. Biol.* 14:18–30.

Tuchman, B. W. 1978. *A distant mirror: The calamitous 14th century.* Alfred A. Knopf, New York.

Turkowski, F. J. 1980. *Carnivora food habits and habitat use in ponderosa pine forests.* Research Paper RM-215. U.S. Forest. Service, Fort Collins, CO.

Turner, J. C., Jr. 1973. Water, energy and electrolyte balance in the desert bighorn sheep, *Ovis canadensis.* Ph.D. dissertation, University of California, Riverside.

U.S. Congress. 1976. National Forest Management Act. Public Law 94-988. U.S. Government Printing Office, Washington, DC.

U.S. Department of Interior. 1966. *Fish, wildlife and pesticides.* U.S. Fish and Wildlife Service, Washington, DC.

Verme, L. J. 1965. Swamp conifer deer yards in northern Michigan. *J. Forest.* 63:523–529.

Verme, L. J. 1977. Assessment of natal mortality in upper Michigan deer. *J. Wildl. Manage.* 41:700–708.

Vilkitis, J. R. 1968. Characteristics of big game violators and extent of their activity in Idaho. M.S. thesis, University of Idaho, Moscow.

Vogl, R. J. 1977. Fire: A destructive menace or natural process. In Cairns, J., K. Dickson, and E. Hendricks. (Eds.), *Recovery and restoration of damaged ecosystems*, pp. 261–289. University of Virginia Press, Charlottesville.

Volterra, V. 1931. Variations and fluctuations of the number of individuals in animal species living together. In R. N. Chapman (Ed.), *Animal ecology*, pp. 409–448. McGraw-Hill, New York.

Wagner, F. H., and L. C. Stoddart. 1972. Influence of coyote predation on black-tailed jackrabbit populations in Utah. *J. Wildl. Manage.* 36:329–343.

Walter, H. 1977. Effects of fire on wildlife communities. In *Proceedings: A symposium on the environmental consequences of fire and fuel management in Mediterranean ecosystems.* Palo Alto, CA, August 1–5, pp. 183–192. General Technical Report WO-3. U.S. Forest Service, Washington, DC.

Ward, A. L. 1976. Elk behavior in relation to timber harvest operations and traffic on the Medicine Bow Range in south-central Wyoming. In J. M. Peek (Chair), *Elk, logging, roads*, pp. 32–43. University of Idaho, Moscow.

Washington Department of Fish and Wildlife. 2008. Washington Department of Fish and Wildlife to begin winter feeding for Mount St. Helens elk. Wildlife Program news release, Olympia, WA, January 16.

Weaver, J. E. 1975. The religion of a scientist. In L. Rosten (Ed.), *Religions of America*, pp. 296–305. Simon and Schuster, New York.

Wesley, D. E., K. L. Knox, and J. G. Nagy. 1970. Energy flux and water kinetics in young pronghorn antelope. *J. Wildl. Manage.* 34:908–912.

Wilkins, K. T. 1982. Highways as barriers to rodent dispersal. *Southwest. Nat.* 27:459–460.

Wilson, E. O. 1975. *Sociobiology: The new synthesis.* Belknap Press, Cambridge, MA.

Wilson, R., K. Ribbeck, and D. Twedt (Eds.). 2007. *Restoration, management and monitoring of forest resources in the Mississippi Alluvial Valley: Recommendations for enhancing wildlife habitat.* Final Report, LMVJV, Forest Resource Conservation Working Group, Vicksburg, MS.

Wishard, L. A. 1989. The red-cockaded woodpecker: A selectively annotated bibliography. *Electronic Green J.* Issue 8. Pennsylvania State University, University Park.

Witter, D. J. 1980. Wildlife values: Applications and information needs in state wildlife management agencies. In W. W. Shaw and E. H. Zube (Eds.), *Wildlife values*, pp. 83–98. Institute Report Number 1, Center for Assessment of Noncommodity Natural Resource Values. University of Arizona, Tucson.

Wynne-Edwards, V. C. 1962. *Animal dispersion in relation to social behavior.* Hafner, New York.

3 Integrating Forestry and Wildlife Management

3.1 INTRODUCTION

Forestland, at least 10% stocked with trees, covers about one-third (32%) of the total land base in the United States or about 731 million acres. Some forestland is found in every state, but the states of North Dakota, South Dakota, Nebraska, and Kansas have less than 5%, while more than 50% of the area in 21 states is forested (U.S. Department of Agriculture [USDA] 1989). Nationally, 482 million acres are *commercial forests* capable of growing economically valuable trees in excess of 20 ft^3. Of the total commercial land, three-fourths is in the eastern United States, equally divided between northern and southern states.

Coniferous forests occur west of the 100th meridian and consist chiefly of large areas of trees extending over the main Rocky Mountain and Pacific Coast ranges. Stands of trees grow along the coast and inland from Alaska to California. Many comparatively small forested tracts are located on ridges and the higher plateaus, intermingled with treeless stretches. They are sometimes widely scattered in great arid districts, especially in parts of the central and southern Rocky Mountains, where large areas of the dry foothills support coniferous woodlands.

In the eastern United States, large unbroken forest areas still exist in the Maritime Provinces, northern New England, northeastern Pennsylvania, and the Appalachian section of the South Atlantic and Gulf states. Southern Florida and Texas have small areas of tropical forests. Forest cover types are identified in detail by color on a 1:7,500,000 scale map for the continental United States (USDA 1967a). A forest type map can be viewed and sections downloaded from the U.S. Geological Survey at http://www.nationalatlas.gov.

3.2 FOREST TYPES

Forest types cover large areas over landscapes. As a result, a type is not continuous and homogeneous but will have gaps and intrusions of patches of other types. Types occur in different areas because some trees have a wider ecological tolerance than others. Types as layers within a state boundary provide different levels of resolution. For the purposes of identifying general type location, the United States has three *geographical areas*: western, central, and eastern (Table 3.1), which is one level of resolution. A second level is *region*, and a third level is *state within a region*. A fourth level is *types within a state*, and a fifth level can be added for a more specific location. An example of a hierarchy in the western United States is

Geographic area	Western United States
Region	Pacific Southwest
State	California
Type	Douglas-fir
Local area	Sierra Mountains

It is important not to assume that the survival and reproduction needs of wildlife can be transferred exactly from one region to another or that a species occurring in a type in one state should also be present in the same type in another state. For example the tassel-eared squirrel occurs in ponderosa pine in Arizona, New Mexico, Colorado, and Utah but does not occur in ponderosa pine in California, Oregon, or Washington. Regionalization of the geographic areas helps to solve this type of problem by grouping states into smaller units for indicating distribution of animals in forest types (Table 3.1).

Each of the 20 National Atlas (USDA 1967a, 2000) forest types (Figures 3.1 and 3.2) can be found in one of the continental 48 states. The number and location of types in each state were interpreted from the atlas map. Types located in the central United States are not indicated on the National Atlas but are included with the eastern types. The central states are contained in a wide belt of plains vegetation that supports some elements of both the western and eastern regions, but the association with forest species is mostly from incursions of the types from either side. A listing for each state shows the distribution of types in states across the continental United States (Table 3.2).

TABLE 3.1
Three Geographical Areas, with 12 Regions, and 48 States with Forest Cover Types

Areas	Regions	States
Western States		
	PNW = Pacific Northwest states	WA, OR
	PSW = Pacific Southwest states	CA
	NRM = Northern Rocky Mountains	ID, MT, WY
	CRM = Central Rocky Mountains	CO, UT, NV
	SRM = Southern Rocky Mountains	AZ, NM
Central States		
	NPS = Northern Plains states	ND, SD, NE, KS
	SPS = Southern Plains states	OK, TX
Eastern States		
	NES = New England states	ME, NH, VT, MA, RI, CT
	NLS = Northern lake states	MN, WI, MI, NY
	MCS = Midcontinent states	OH, IN, KY, IL, IA, MO, PA, WV
	SAS = South Atlantic states	NC, SC, GA, FL, VA, MD, DE, NJ
	SCS = South Central states	TN, AL, MS, AR, LA

FIGURE 3.1 Forest types in the eastern United States.

The two most dominant types (based on size of area occupied) in each state are shown with the other types that are present. Thirty-eight of 48 states have a national forest where examples of the types can be found (Table 3.3). The descriptions of forest types that follow are short summaries containing introductory information. For each type, a checklist of common forest wildlife species is provided, but an individual species may not be in every state where the type occurs.

3.2.1 EASTERN FOREST TYPES

3.2.1.1 White-red-jack pine (Type 10)

The white-red-jack pine type (Type 10) occurs on smooth-to-irregular plains and tablelands of the northern lake states and provinces and parts of New York, New

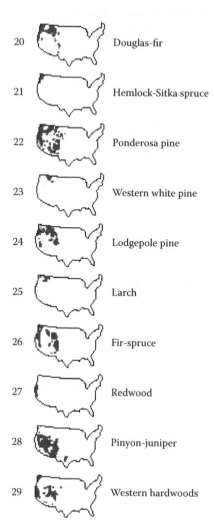

FIGURE 3.2 Forest types in the western United States.

England, Quebec, and the Maritime Provinces. Eastern white pine, red pine, and jack pine are the major components. Associates include eastern hemlock, aspens, northern white cedar, maples, and birches. Precipitation averages 25–45 in, distributed evenly throughout the year. More than 50% of the area is gently sloping. Timber harvesting and recreation are the main land use activities.

3.2.1.1.1 Common Wildlife Species Checklist

Common wildlife species are white-tailed deer, black bear, moose, coyote, bobcat, snowshoe hare, great horned owl, spruce grouse, ruffed grouse, porcupine, blackburnian warbler, black-throated green warbler, whip-poor-will, crested flycatcher, wood pewee, white-breasted nuthatch, veery, pileated woodpecker, hairy woodpecker, downy woodpecker, blue jay, chickadee, red-eyed vireo, black-and-white

TABLE 3.2
Dominant Forest Types in the 48 Continental States

State	Scode	% Forests	Dominant Types	Other Types
Alabama	AL	66	13, 14	12, 15, 16
Arkansas	AR	55	13, 16	14, 15
Arizona	AZ	25	22, 28	26
California	CA	40	20, 27	24, 26, 29
Colorado	CO	34	24, 26	20, 22, 28
Connecticut[a]	CT	60	15	10, 17, 18
Delaware[a]	DE	31	13	14, 15
Florida	FL	50	12	15, 16
Georgia	GA	68	12, 13	14, 15, 16
Iowa[a]	IA	04	15, 17	15
Idaho	ID	41	20, 23	22, 24, 25, 26
Illinois	IL	11	15, 17	16
Indiana	IN	17	15, 17	16, 18
Kansas[a]	KS	02	15, 17	
Kentucky	KY	48	14, 15	16, 17
Louisiana	LA	51	13, 16	12, 14
Massachusetts[a]	MA	59	15, 18	10, 13, 17
Maryland[a]	MD	41	13, 14	15
Maine	ME	90	10, 11	15, 18
Michigan	MI	53	18, 19	10, 15, 17
Minnesota	MN	36	10, 19	11, 15, 17, 18
Missouri	MO	29	14, 15	16, 17
Mississippi	MS	56	13, 16	12, 14, 15
Montana	MT	24	24, 25	20, 22, 26, 29
North Carolina	NC	64	13, 15	10, 12, 14, 16
North Dakota[a]	ND	01	22, 19	15, 17, 29
Nebraska	NE	02	17, 22	
New Hampshire	NH	87	10, 18	11, 15
New Jersey[a]	NJ	40	13, 15	14, 17
New Mexico	NM	23	22, 28	20
Nevada	NV	11	28	22, 29
New York	NY	57	15, 18	10, 11, 17
Ohio	OH	25	15, 18	17
Oklahoma[a]	OK	19	14, 15	13, 16
Oregon	OR	49	20, 21	22, 24, 25, 26, 29
Pennsylvania	PA	62	15, 18	15, 18
Rhode Island[a]	RI	61	15	10, 17
South Carolina	SC	63	12, 13	14, 15, 16
South Dakota	SD	03	17, 22	29
Tennessee	TN	50	14, 15	10, 13, 16

(continued on next page)

TABLE 3.2 (continued)
Dominant Forest Types in the 48 Continental States

State	Scode	% Forests	Dominant Types	Other Types
Texas	TX	14	13, 15	12, 14, 16, 28
Utah	UT	30	26, 28	20, 22, 24, 29
Virginia	VA	64	13, 15	10, 14
Vermont	VT	76	11, 18	10, 11, 18
Washington	WA	55	20, 21	22, 23, 24, 25, 26, 29
Wisconsin	WI	43	15, 19	10, 11, 17, 18
West Virginia	WV	76	11, 15	13, 18
Wyoming	WY	16	24, 26	20, 22

[a] States without a national forest.

warbler, ovenbird, redstart, black-throated blue warbler, hermit thrush, magnolia warbler, Canada warbler, yellow-bellied sapsucker, olive-sided flycatcher, red-breasted nuthatch, brown creeper, winter wren, blue-headed vireo, and myrtle warbler. The Kirkland's warbler occurs in limited areas.

3.2.1.2   Spruce-fir (Type 11)

Flat plains and tablelands in the lake states and provinces, New England states, and the Maritime Provinces at high elevations in the Appalachian Mountains are occupied by the spruce-fir type (Type 11). Red spruce, white spruce, and balsam fir are associated with northern white cedar, eastern hemlock, eastern white pine, and tamarack. Hardwoods include maples, birches, and aspens. Normal precipitation is 30–40 in. The type occurs at high elevations in the Appalachian Mountains as far south as West Virginia. Relief is less than 500 ft and often below 100 ft. Timber harvesting and recreation are the main land use activities.

3.2.1.2.1   Common Wildlife Species Checklist

Common wildlife species are moose, woodland caribou, Canada lynx, American marten, black bear, long-tailed weasel, white-footed mouse, white-tailed deer, ruffed grouse, wild turkey, spruce grouse, olive-backed thrush, magnolia warbler, Cape May warbler, myrtle warbler, bay-breasted warbler, and white-throated sparrow.

3.2.1.3   Longleaf-slash pine (Type 12)

The longleaf-slash pine type (Type 12) is restricted to flat and irregular southern Gulf coastal plains of the United States where local relief is less than 300 ft. Major species include longleaf and slash pine. Associated species include shortleaf, loblolly, and Virginia pine, with a mixture of oaks and hickories. Annual precipitation ranges from 40 to 60 in. Wire grasses are the main herbaceous plants, but other grasses, such as panicums, paspalums, and dropseed, are common. Gallberry, saw palmetto, wax myrtle, and shining sumac are prominent shrubs. Forage yields can be sustained by periodically burning the range to improve quality and palatability.

TABLE 3.3
National Forests in 38 States

State	National Forests
Alabama	Conecuh, Talladega, Tuskegee, Bankhead
Arkansas	Ouachita, Ozark-St. Francis
Arizona	Apache-Sitgreaves, Coconino, Coronado, Kaibab, Prescott, Tonto
California	Angeles, Cleveland, Eldorado, Inyo, Klamath, Lassen, Los Padres, Mendocino, Modoc, Plumas, San Bernardino, Sequoia, Shasta-Trinity, Sierra, Six Rivers, Stanislaus, Tahoe
Colorado	Arapaho, Grand Mesa, Gunison, Pike, Rio Grande, Roosevelt Routt, San Isabel, San Juan, Uncompahgre, White River
Florida	Apalachicola, Ocala, Osceola
Georgia	Oconee, Chattahoochee
Idaho	Panhandle, Boise, Caribou, Challis, Clearwater, Nez Perce, Payette, Salmon, Sawtooth, Targhee
llinois	Shawnee
Indiana	Hoosier
Kentucky	Daniel Boone
Louisiana	Kisatchie
Maine	White Mountains
Michigan	Hiawatha, Huron-Manistee, Ottawa
Minnesota	Chippewa, Superior
Mississippi	Bienville, Delta, Desoto, Holy Springs, Homochitto, Tombigee
Missouri	Mark Twain
Montana	Beaverhead-Deerlodge, Bitterroot, Custer, Flathead, Gallatin, Helena, Kootenai, Lewis and Clark, Lolo
Nebraska	Nebraska
Nevada	Humboldt-Toiyabe
New Hampshire	White Mountain
New Mexico	Carson, Cibola, Santa Fe, Lincoln, Gila
New York	Finger Lakes
North Carolina	Croatan, Nantahala, Pisgah, Uwharrie
Ohio	Wayne
Oregon	Deschutes, Fremont, Malheur, Mount Hood, Ochoco, Rouge River, Siskiyou, Siuslaw, Umatilla, Umpqua, Wallowa-Whitman, Willamette, Winema
Pennsylvania	Allegheny
South Carolina	Francis Marion-Sumter
South Dakota	Black Hills
Tennessee	Cherokee
Texas	Angelina, Davy Crockett, Sabine, Sam Houston
Utah	Ashley, Dixie, Fishlake, Manti-LaSal, Uninta, Wasatch-Cache
Vermont	Green Mountain
Virginia	Jefferson, George Washington

(continued on next page)

TABLE 3.3 (continued)
National Forests in 38 States

State	National Forests
Washington	Colville, Gifford Pinchot, Mount Baker-Snoqualmie, Okanogan, Olympic, Wenatchee
West Virginia	Monogahela
Wisconsin	Chequamegon, Nicolet
Wyoming	Bighorn, Bridger Teton, Medicino Bow, Shoshone

3.2.1.3.1 *Common Wildlife Species Checklist*

Common wildlife species are white-tailed deer, northern raccoon, Virginia opossum, gray squirrel, eastern cottontail, bobwhite quail, red-cockaded woodpecker, and black bear. Cougar (Florida panther) may be seen but are rare.

3.2.1.4 Loblolly-shortleaf pine (Type 13)

The loblolly-shortleaf pine type (Type 13) generally occurs on irregular Gulf coastal plains and the Piedmont, where local relief is 300–1,000 ft. This type is of the greatest extent in the South and Southeast. The major species are loblolly and shortleaf pine in association with oaks, hickories, and maples. Annual precipitation is 40–60 in evenly distributed with a midsummer peak. About 60–70% of the land use is in forestry, mostly farm woodlots. The main grasses are bluegrass and panicums. Dogwood, viburnum, haw, blueberry, American beautyberry, yaupon, and woody vines are common.

3.2.1.4.1 *Common Wildlife Species Checklist*

Common wildlife species are white-tailed deer, eastern cottontail, fox squirrel, northern raccoon, red fox, wild turkey, bobwhite quail, mourning dove, pine warbler, northern cardinal, summer tanager, Carolina wren, ruby-throated hummingbird, blue jay, hooded warbler, eastern towhee, tufted titmouse, and red-cockaded woodpecker.

3.2.1.5 Oak-pine (Type 14)

The oak-pine type (Type 14) occurs on diverse land forms from the southernmost ridges and valleys of the Appalachians westward across the coastal plains and north into the Ozark plateaus and Ouachita provinces. White oak, northern red oak, scarlet oak, black oak, chestnut oak, with loblolly pine, shortleaf pine, and Virginia pine are the major species. Precipitation is relatively high, averaging 40–55 in.

3.2.1.5.1 *Common Wildlife Species Checklist*

Common wildlife species are white-tailed deer, fox squirrel, eastern cottontail, mourning dove, bobwhite quail, wild turkey, ruffed grouse, northern cardinal, tufted titmouse, wood thrush, summer tanager, red-eyed vireo, blue-gray gnatcatcher, hooded warbler, Carolina wren, eastern box turtle, common garter snake, and timber rattlesnake.

3.2.1.6   Oak-hickory (Type 15)

Oak-hickory (Type 15) is the most extensive type in the United States and occurs in the East, where conditions are more mesophytic than the surrounding areas. White oak, northern red and black oak, with associates of pignut, shagbark, mockernut, and bitternut hickory are the major species in the type. The forest type is best developed in the Ozarks. Annual precipitation in the type averages between 35 and 45 inches. Oak-hickory varies from open to closed woods with a strong-to-weak understory of shrubs, vines, and herbaceous plants. Major shrubs are blueberry, viburnum, dogwood, rhododendron, and sumac. Vines include woodbine, grape, poison ivy, greenbrier, and blackberry. Herbaceous plants are sedge, panicum, bluestem, lespedeza, tickclover, goldenrod, pussytoes, and aster.

3.2.1.6.1   Common Wildlife Species Checklist

Common wildlife species are white-tailed deer, black bear, bobcat, gray fox, northern raccoon, gray squirrel, fox squirrel, eastern chipmunk, white-footed mouse, pine vole, short-tailed shrew, cotton mouse, wild turkey, ruffed grouse, bobwhite quail, mourning dove, northern cardinal, tufted titmouse, wood thrush, summer tanager, red-eyed vireo, blue-gray gnatcatcher, hooded warbler, Carolina wren, eastern box turtle, common garter snake, and timber rattlesnake.

3.2.1.7   Oak-gum-cypress (Type 16)

The oak-gum-cypress type (Type 16) is characterized by the vegetation of the Mississippi Valley and other bottomlands in every southern state, the cypress savanna west of the Everglades in Florida, the mangrove swamps south of the Everglades, and the eastern coast of Florida, Georgia, and the Carolinas. Blackgum, sweetgum, bald cypress, nuttall, shumard, overcup, and cherrybark oak are the major species in the type. Other hardwoods, such as willows, maples, sycamore, cottonwoods, and beech, are present depending on location. Annual precipitation varies from 60 in (in Florida) to 35 in (in the northern extremity of the type in Indiana).

3.2.1.7.1   Common Wildlife Species Checklist

Common wildlife species are white-tailed deer, black bear, bobcat, gray fox, fox squirrel, northern raccoon, Virginia opossum, striped skunk, eastern cottontail, swamp rabbit, wild turkey, Bachman's warbler, bald eagle, and mangrove cuckoo.

3.2.1.8   Elm-ash-cottonwood (Type 17)

The elm-ash-cottonwood type (Type 17) occurs in narrow belts along major streams or scattered areas of dry swamps largely on the lower terraces and flood plains of the Mississippi, Missouri, Platte, Kansas, and Ohio Rivers from the Dakotas, Minnesota, and Ohio south through Kansas and Missouri. Cottonwood is the dominant species in the type, with associates of green ash, white ash, American elm, and black willow. Precipitation varies from 10 in near the foothills of the Rocky Mountains to 50 in in the southern and northeastern areas. Vegetation is of low-to-tall broad-leaved deciduous trees, varying from open to dense and often accompanied by vines.

3.2.1.8.1   Common Wildlife Species Checklist

Common wildlife species are eastern cottontail, bobwhite quail, white-tailed deer, northern raccoon, red fox, coyote, striped skunk, spotted skunk, meadow jumping mouse, fox squirrel, sharp-tailed grouse, ruffed grouse, catbird, goldfinch, yellow-billed cuckoo, indigo bunting, northern cardinal, lark sparrow, mockingbird, common crow, blue jay, American robin, ruby-throated hummingbird, and Cooper's hawk.

3.2.1.9   Maple-beech-birch (Type 18)

The maple-beech-birch type (Type 18) is best developed in the New England states and Maritime Provinces. Sugar maple, American beech, and yellow birch are the dominant tree species. Associates are red maple, eastern hemlock, white ash, black cherry, sweet birch, and northern red oak. Typically, the type occurs on open high hills and low mountains. Annual precipitation ranges from 40 to 48 in.

3.2.1.9.1   Common Wildlife Species Checklist

Common wildlife species are white-tailed deer, black bear, red fox, gray fox, bobcat, ruffed grouse, bobwhite quail, ovenbird, red-eyed vireo, hermit thrush, scarlet tanager, blue jay, black-capped chickadee, wood pewee, and magnolia warbler.

3.2.1.10   Aspen-birch (Type 19)

The aspen-birch type (Type 19) is found within the Great Lakes region. Quaking aspen, bigtooth aspen, balsam poplar, paper birch, and gray birch are the dominant species. Balsam fir and maples are also common. The type is closely associated with moraines and outwash plains of recent glacial origin. Annual precipitation for the type is between 30 and 35 in.

3.2.1.10.1   Common Wildlife Species Checklist

Common wildlife species are white-tailed deer, black bear, coyote, bobcat, great horned owl, ruffed grouse, tufted titmouse, blue jay, hairy woodpecker, downy woodpecker, wood thrush, eastern wood pewee, goldfinch, catbird, and red-eyed vireo.

3.2.2   Western Forest Types

3.2.2.1   Douglas-fir (Type 20)

Just south and west of the hemlock-spruce forest and inland from the coast of British Columbia, Washington, and Oregon extending south to northern California is the Douglas-fir forest (Type 20). This forest contains at least 50% Douglas-fir with varying amounts of western hemlock, western red cedar, and redwood. Pacific yew and bigleaf maple are common understory species. The type constitutes one of the largest blocks of timber in the West. In the Rocky Mountains, the type is developed best in Idaho, Montana, British Columbia, and Alberta, but its range extends southward to Utah, Colorado, New Mexico, and Arizona. Associated species are lodgepole pine, western larch, grand fir, and quaking aspen. The Douglas-fir forest is the most widespread type in the northern Rocky Mountains, but southward it is scattered.

Precipitation is 40–80 in. in the extreme west to 20–30 in. in the interior. Common shrubs are rock spirea, filbert, blueberry, snowberry, barberry currant, blackberry, ninebark, and rose.

3.2.2.1.1 *Common Wildlife Species Checklist*

Common wildlife species are elk, mule deer, black bear, brown bear (grizzly), moose, blue grouse, ruffed grouse, cougar, bobcat, American marten, bushy-tailed wood rat, chestnut-backed chickadee, red-breasted nuthatch, gray jay, spotted owl, and Steller's jay.

3.2.2.2 Hemlock-Sitka spruce (Type 21)

Southward along the coast of British Columbia, Washington, Oregon, and northern California, western hemlock and Sitka spruce (Type 21) are the dominant species, but Douglas-fir and western red cedar are common associates. Red alder and black cottonwood are also present. Sitka spruce generally exceeds 200 ft in height and can attain ages of more than 1,000 yr. Bigleaf maple is the only major hardwood component. Precipitation ranges from 60 to 115 in annually. The type probably has the most productive understory of any coniferous forest in the United States.

3.2.2.2.1 *Common Wildlife Species Checklist*

Common wildlife species are elk, mule deer, black bear, moose, red-tailed hawk, western screech owl, pygmy owl, great horned owl, cougar, bobcat, gray wolf, American marten, western spotted skunk, deer mouse, Douglas squirrel, bushy-tailed wood rat, Townsend's chipmunk, coast mole, red crossbill, chestnut-backed chickadee, red-breasted nuthatch, common raven, gray jay, Steller's jay, hermit warbler, western wood pewee, pine siskin, blue grouse, and ruffed grouse.

3.2.2.3 Ponderosa pine (Type 22)

Extending south through central British Columbia, Washington, and Oregon and into California is the narrow ponderosa pine type (Type 22). Associated with ponderosa pine in the northern areas are Douglas-fir, lodgepole pine, grand fir, and western larch. South into California, common associates are white fir, Jeffrey pine, and incense cedar. Ponderosa pine occupies warm, dry sites in the Rockies, with precipitation from 15 to 20 in. Its range extends from eastern British Columbia, Washington, and Oregon, into Idaho, Montana, Wyoming, and South Dakota and south through Colorado, New Mexico, and Arizona. At higher elevations, Douglas-fir is associated with ponderosa pine. In the northern part of its range, lodgepole pine, blue spruce, and quaking aspen are common species.

3.2.2.3.1 *Common Wildlife Species Checklist*

Common wildlife species are elk, mule deer, cougar, coyote, bushy-tailed wood rat, white-footed mouse, bobcat, rock squirrel, eastern cottontail, porcupine, golden-mantled ground squirrel, tassel-eared squirrel, pygmy nuthatch, long-crested jay, sharp-shinned hawk, Rocky Mountain nuthatch, mountain chickadee, Cassin's purple finch, red-shafted flicker, red-backed junco, northern goshawk, red-tailed hawk, chestnut-backed bluebird, Audubon's warbler, Natalie's sapsucker, western chipping sparrow, great horned owl, and band-tailed pigeon.

3.2.2.4 Western white pine (Type 23)

The western white pine type (Type 23) occurs in the high mountains of the northern Rocky Mountains of western Alberta, Montana, and northern Idaho. It is also present in scattered areas in the Cascade Mountains of Oregon and Washington. White pine is the dominant species in combination with sugar pine. Western red cedar is an associated species. Annual rainfall in the type is from 20 to 30 in.

3.2.2.4.1   Common Wildlife Species Checklist

Common wildlife species are elk, mule deer, snowshoe hare, long-tailed weasel, American marten, coyote, bobcat, black bear, ruffed grouse, chestnut-backed chickadee, red-breasted nuthatch, and Swainson's thrush.

3.2.2.5 Lodgepole pine (Type 24)

East of the ponderosa pine type in California is the lodgepole pine type (Type 24). It extends northward as a disjunct population primarily in Oregon. The major associated species are red fir and mountain hemlock. The type occurs in pure stands in British Columbia, Idaho, and Montana, but it ranges from Oregon and Wyoming south into Colorado. Common associated species are Douglas-fir, Engelmann spruce, subalpine fir, and aspens. Lodgepole pine as a pioneer species gets established following fire, but it can also form a climax forest. Annual rainfall is from 20 to 50 in.

3.2.2.5.1   Common Wildlife Species Checklist

Common wildlife species are elk, mule deer, black bear, brown bear (grizzly), moose, blue grouse, ruffed grouse, cougar, bobcat, American marten, bushy-tailed wood rat, chestnut-backed chickadee, red-breasted nuthatch, gray jay, spotted owl, and Steller's jay.

3.2.2.6 Larch (Type 25)

Located in the high mountains of Alberta, eastern Oregon, northern Idaho, and western Montana is the larch forest type (Type 25). Western larch is the dominant species in association with white pine. Also, Douglas-fir and grand fir may be present. Annual precipitation is 20–30 in. in eastern Oregon and 20–50 in. in the northern Rockies.

3.2.2.6.1   Common Wildlife Species Checklist

Common wildlife species are elk, mule deer, black bear, brown bear (grizzly), moose, blue grouse, ruffed grouse, cougar, bobcat, marten, bushy-tailed wood rat, chestnut-backed chickadee, red-breasted nuthatch, gray jay, spotted owl, and Steller's jay.

3.2.2.7 Fir-spruce (Type 26)

Adjacent to the ponderosa pine but at higher elevations is another long, narrow band of vegetation, fir-spruce (Type 26), comprised of silver fir, subalpine fir, red fir, white fir, and Engelmann spruce, sometimes in combination with mountain hemlock. In the Southwest, the type is developed best in Colorado, but it also occurs in northwest Wyoming and in disjunct areas in western Montana and eastern Idaho. Fir-spruce occurs in small areas at high elevation in Arizona and New Mexico; species in the type are subalpine and white fir with Engelmann spruce, blue spruce, and aspen. Annual precipitation ranges from 22 in. in the Rocky Mountains to 75 in. in the Sierras.

3.2.2.7.1   Common Wildlife Species Checklist
Common wildlife species are elk, mule deer, white-tailed deer, wolverine, Canada lynx, black bear, cougar, coyote, gray wolf, brown bear (grizzly), bighorn sheep, blue grouse, spruce grouse, ruffed grouse, porcupine, American beaver, snowshoe hare, flying squirrel, pocket gopher, western bluebird, American robin, and Steller's jay.

3.2.2.8   Redwood (Type 27)
The redwood type (Type 27) occurs only in the low coastal mountains of northern California and the southwestern portion of Oregon. Redwoods are the tallest trees in the world, often reaching 300 ft. Douglas-fir, western hemlock, and Sitka spruce are common associates. Annual precipitation is approximately 60 in. Hardwood species found in the redwood type are bigleaf maple, red alder, and willows.

3.2.2.8.1   Common Wildlife Species Checklist
Common wildlife species are mule deer, elk, cougar, bobcat, black bear, red-tailed hawk, western screech owl, great horned owl, sharp-shinned hawk, Cooper's hawk, band-tailed pigeon, blue grouse, pileated woodpecker, gray jay, brown creeper, hermit warbler, and red crossbill.

3.2.2.9   Pinyon-juniper (Type 28)
The pinyon-juniper type (Type 28) occurs in the basin and range province of Utah, Nevada, southern Idaho, and southeast Oregon. Large areas of pinyon-juniper occur in Arizona and New Mexico. Singleleaf and Colorado pinyon with Utah, alligator, and one-seed juniper are the main species. The name *pygmy forest* characterizes the pinyon and juniper woodland. The trees occur as dense-to-open woodland and savannah woodland. Annual precipitation is about 10 in.

3.2.2.9.1   Common Wildlife Species Checklist
Common wildlife species are mule deer, cougar, coyote, bobcat, elk, white-footed mouse, cliff chipmunk, black-tailed jackrabbit, eastern cottontail, rock squirrel, North American porcupine, gray fox, ring-tailed cat, spotted skunk, gray titmouse, Woodhouse's jay, western red-tailed hawk, golden eagle, red-shafted flicker, pinyon jay, lead-colored bushtit, rock wren, chipping sparrow, night hawk, black-throated gray warbler, northern cliff swallow, western lark sparrow, Rocky Mountain grosbeak, desert sparrow, mourning dove, pink-sided junco, Shufeldt's junco, gray-headed junco, red-backed junco, Rocky Mountain nuthatch, mountain bluebird, western robin, long-crested jay, wild turkey, horned lizard, sagebrush swift, collared lizard, and Great Basin rattlesnake.

3.2.2.10   Western hardwoods (Type 29)

The western hardwoods type (Type 29) occurs in California and Oregon and in the Rocky Mountain regions with rainy winters and dry summers. Precipitation varies from 40 to 100 in. Vegetation consists mostly of Oregon white oak, Coulter and digger pine, blue oak, canyon live oak, and interior live oak. In some areas, aspen and other conifers are present with an understory of shrubs, grasses, and forbs.

3.2.2.10.1   Common Wildlife Species Checklist

Common wildlife species are mule deer, coyote, bobcat, golden eagle, red-tailed hawk, San Joaquin kit fox, California quail, mountain quail, striped skunk, kangaroo rat, pocket gopher, and western gray squirrel.

Wildlife species listed under the three U.S. regions (west, central, and east) are the first level of resolution in the FAAWN (forest attributes and wildlife needs) data model discussed in Chapter 5. A second level lists animals in states within a region and was developed by overlaying the geographic range of an animal on individual state boundaries as documented in major field guides and regional publications. The third level of resolution is the potential association of animals by forest type in a state. In the FAAWN data model, there are over 63,000 listings for which habitat is *hypothesized* to be available for a given species. Presence has been validated for species in some types in states, but others are considered candidates. When a forest type is identified as containing habitat for a given animal, that only means that there are certain spots within the type that are habitat and not the total type expanse. Other levels of resolution beyond type can be added to refine an association, such as the Sierras in California, Cascades in Oregon, or Appalachians in North Carolina.

The scientific name, common name, and life form of all plants and animals mentioned in this textbook are listed in the animal and plant tables in the FAAWN data model on the enclosed computer disk. Authorities for names used are as follows: eastern amphibians and reptiles, Conant and Collins (1998); western amphibians and birds, Stebbins (2003); mammals, Reid (2006); eastern birds, Peterson (1980); western birds, Peterson (1990); eastern trees, Petrides (1988); and western trees, Petrides (1992).

3.3   FOREST MANAGEMENT

Wildlife management has often been in conflict with other resource management practices and uses, but the two principal areas of contention, in both the past and present, are the harvesting of trees and the grazing of forested areas by livestock. The issues of the past are still with us but at a more refined level. In the 1930s, the *Journal of Forestry* included articles discussing the relationship of forestry and wildlife (Chapman 1936; Gabrielson 1936; Cahalane 1939) and in 1940 that of timber versus wildlife (Davenport 1940). In recent years, that journal included articles on biological diversity (Blockstein 1990), wildlife management (Wood 1990), the new focus for wildlife management (Giles and Nielson 1990), the redesigned forest (Lee 1990), and spotted owls (Fox and Rhea 1989).

The concerns of the past will continue to be the basis of conflict in the future for issues such as the cutting of old-growth forests, snags, and riparian trees. These issues have created the controversies involving the spotted owl, cavity-nesting birds, bald eagles, and other species classified as threatened or endangered. Yet, timber harvesting to provide wood for human use while maintaining a sustainable forest ecosystem can be complementary if the objectives for a management unit are clearly defined to include both. The importance of forest ecosystems to wildlife is best understood from the statistic that 1,578 species in the United States are found in 48 states, and 1,101 species are believed to be associated with 20 forest types. These species form the core relation and the emphasis of the FAAWN data model.

3.3.1 SILVICULTURAL SYSTEMS

In the forest ecosystem, timber harvesting is the primary means available to managers to set back succession for producing food and cover for some wildlife species now and for the future. *Silviculture* is the art and science of controlling the establishment, growth, composition, health, and quality of forests and woodlands to meet the diverse needs and values of landowners and society on a sustainable basis (Helms 1998). Given a set of goals, which remain constant over a sufficiently long period of time, the silviculturist can prescribe treatments for stands of trees to meet ecosystem management goals. In fact, the definition of silviculture includes this emphasis.

Silvicultural systems can be used not only to produce timber but also, with some alternations from the traditional methods, to meet multiresource objectives of wildlife, watershed, recreation, visual quality, and range management objectives. Ideally, silviculture begins with the regeneration of a forest stand following harvesting. However, silvicultural treatments can start anytime in the life cycle of a forest stand. Once a stand has been regenerated and has progressed to the sapling, pole, or saw timber stage, thinning can be used to influence subsequent tree growth and arrangement. By convention, a silvicultural system is named after the process used to regenerate a stand. Silviculture is an applied science and a working hypothesis influenced by decisions from the best information available. As a result, the practice of silviculture does not have strict guidelines, and foresters managing a given stand or forest need to consider that trees are just one part of a functioning ecosystem.

3.3.2 TREE SPECIES TOLERANCE

An understanding of light tolerance of a species is critical in selecting a method to regenerate and manage a forest. For example, loblolly pine cannot grow in stands of larger and older trees that have a high degree of canopy closure. To establish the species, it is necessary to cut all the trees in a designated area to remove the shade cast by overstory trees. Northern hardwoods, such as sugar maple and yellow birch, however, grow under a relatively closed canopy and thus can be regenerated in shade.

Every tree species has a different requirement for the amount of light needed for vigorous, healthy growth. Species needing large amounts of sunlight (e.g., ponderosa pine, yellow poplar, and Virginia pine) are categorized as *intolerant*; that is, they cannot grow in shade. Conversely, sugar maple, Engelmann spruce, redwood, and western hemlock can survive and grow in reduced sunlight and thus are termed *tolerant* species (Baker 1949). Ranging between the two categories is a spectrum of species that make up an *intermediate class*, for example, Douglas-fir, white oak, and sycamore. The tolerance of species is related to successional development; intolerant species are the first to germinate or sprout and grow following disturbance, while the tolerant species are established later in the tree replacement process that will ultimately become the climax stage.

3.3.3 Even- and Uneven-Aged Stands

Tree species requiring considerable sunlight to germinate and grow benefit from even-aged management. *Even-aged* stands have trees in a single age class in which the range of tree ages is usually +20% of the rotation. Three regeneration methods are even aged (clearcutting, shelterwood, and seed tree), and one is uneven aged (selection). The shelterwood method could be designed to maintain and regenerate a stand with two age classes (Helms 1998).

Even-aged management is directed at establishing a cohort of trees that can be harvested in the future. At the time of harvesting, all the trees will be in the same age or size class. Uneven-aged stands, produced by single-tree or group selection, have trees of three or more distinct age classes, either intimately mixed or in small groups (Helms 1998). Uneven-aged forests can also be managed as small units of even-aged stands. There are differences between even- and uneven-aged management, such as species composition and frequency and intensity of harvest on a particular acre, but the fundamental difference between the two is the arrangement of the growing stock (Figure 3.3) created by the different regeneration methods (Figure 3.4).

3.3.4 Clearcutting

The harvesting of all trees in a given area at one time is *clearcutting*. The objective is to create an even-aged stand. A new stand regenerates naturally by seed from the surrounding trees, from root or stump sprouts, by artificially planting nursery-grown seedlings, or by direct seeding. The size of a clearcut is dictated by the requirements of the tree species, economics, state and federal laws, and constraints imposed to protect other resources (e.g., scenic beauty and wildlife). Sizes of clearcuts can range from a small stand of 10 ac to over 1,000 ac depending on the tree species. No matter what the size, the primary objective is to regenerate the stand in a short time interval by cutting strips, blocks (Figure 3.5), or designing for the natural landscape (Figure 3.6).

Clearcutting as an even-aged regeneration method became controversial during hearings of the Senate Committee on Interior and Insular Affairs in April 1971 (Brooks 1971). The controversy centered on the size of clearcuts amid cries of "rape of the national forests" in West Virginia and Wyoming (2,000 ac on the Bridger

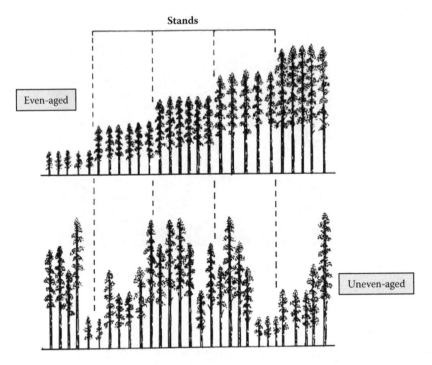

FIGURE 3.3 Visual representation of even-aged and uneven-aged stands in the arrangement of growing stock.

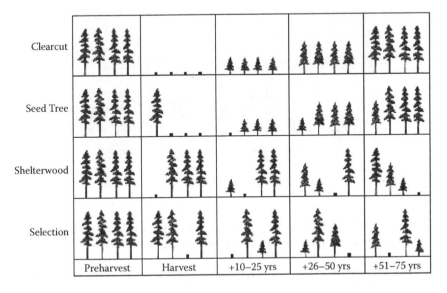

FIGURE 3.4 Diagrammatic concept of regeneration methods.

FIGURE 3.5   Clearcutting in blocks in the Pacific Northwest. (USDA Forest Service photograph.)

FIGURE 3.6   Clearcutting to achieve a natural landscape pattern with edge. (USDA Forest Service photograph.)

Teton National Forest in Wyoming). The question asked by the associate chief of the Forest Service was appropriate for that time and is appropriate now:

> When the Lord looked out on his work and found that it was good, He enjoyed a pre-rogative not given to mortal man. Mortals must always ask questions of value—good for what and for whom. These questions apply to analyzing the benefits of clearcutting for wildlife. (Resler 1972, p. 250)

One response to the "for what and for whom" was provided by Edward Hooven:

> What is apparent is that clearcutting creates openings distributed in a patchwork fashion. Because the openings occur at different time intervals, they offer a continuing range of optimum habitat for wildlife of all kinds. The combination of forest, edge, and clearcutting supplies more varied species of animals, from insects to big game, than does any one of the three situations alone. (1973, p. 211)

As a result of the concern by the public and the need to resolve a critical management issue involving public lands, the National Forest Management Act (NFMA), after a long period of public and congressional debate (Brooks 1971), was passed in 1976. Regulations implementing the act put the Forest Service under more congressional scrutiny through a detailed planning process that is still enforced today.

3.3.5 SHELTERWOOD

If trees on a site are removed over time with the intent of creating an even-aged stand in a series of cuts, the method is referred to as *shelterwood* harvesting (Figure 3.7). Mature trees that remain, following the initial cut, provide both seed and shelter for seeds to germinate and for seedlings to get established. Several intermediate cuts are common, but the last one always removes the remaining shelter trees. The first entry is a preparatory cut to open the forest canopy by about 50% to stimulate seed production and provide space for seedlings. A second cut, the seed cut, can be made to remove another 25% of the original stand. Seed cuts can be bypassed if adequate seedling establishment occurred following the preparatory cut. A third or final cut is made to remove the remaining shelter trees after seedlings are established. Shelterwood cuts can be in strips, patches, blocks, or a geometric shape along the landscape.

FIGURE 3.7   Leave trees in a shelterwood cut. (USDA Forest Service photograph.)

3.3.6 SEED TREE

Removing almost all the trees but leaving a few scattered mature trees of good genetic stock to produce seed to regenerate the harvested area is *seed tree* harvesting (Figure 3.4). The seed source can be a single tree or a small group of two or three trees. After adequate regeneration has been established, the remaining seed trees are removed. As with clearcutting and shelterwood harvesting, the seed tree cut is used for regenerating intolerant species of even-aged stands.

3.3.7 SELECTION

Removing single trees or small groups (*group selection*) at each cutting is *selection harvesting* (Figure 3.4). The objective of this method is to create or maintain an uneven-aged stand in which regeneration advances continuously. If the groups of trees harvested are large, the result resembles a small clearcut. Trees selected for cutting are of all ages and diameters, including the oldest or largest in the stand. In the selection method, cuttings are made more frequently in a stand than in the other three harvest methods. The selection method is designed to duplicate forest succession where individual trees die and a gap is created where new regeneration can be established. The gaps created by dead trees contribute to diversity within a stand.

3.3.8 OLD FOREST CONDITIONS

The NFMA of 1976 focused attention on the need to maintain and provide for diverse plant and animal communities, including old-growth forests (Figure 3.8). Forest communities aged 175–1,000 years provide habitat for certain associated wildlife species, such as the spotted owl. Under the law, all the values found in these old

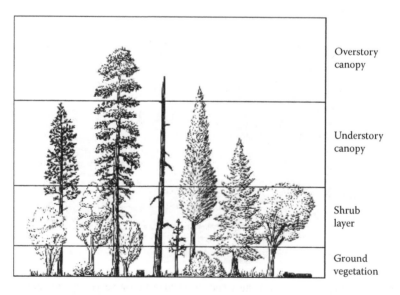

Overstory canopy

Understory canopy

Shrub layer

Ground vegetation

FIGURE 3.8 Diversity of a stand in an old forest condition.

forest communities must be considered, not simply their value for timber production. The old-growth forests in the Pacific Northwest became the center of attention in the 1980s due to the impact of this legislation on the increased demand for wood products that affected the habitat of the spotted owl.

In response to a lawsuit by the Seattle Audubon Society, the federal court issued an injunction against timber sales until the Forest Service could develop a management plan. This resulted in the formation of a scientific analysis team to respond to questions raised by Judge Dwyer. Later, President Clinton hosted a forest conference in Portland, Oregon, that included all the interested parties. The president asked that a forest management plan be developed that would meet certain goals. The management plan was completed in April 1994, but the controversies continue.

3.3.9 THINNING

Harvesting selected trees before a final or regeneration cut is *thinning*. If the trees are usable, then it is a *commercial thinning*. A stipulated percentage of trees is removed from a stand to improve the growth or form of the remaining trees. If the cut is made to remove trees damaged by fire, insects, or disease, it is a *salvage cut*. The number and intensity of thinnings made before the final harvest varies according to management objectives and how well the stand responded to prior thinning.

3.3.10 SPACING OF TREES

Thinning dense stands of unevenly spaced trees promotes diameter growth and increases the quality of wood, but it can have a negative effect on some wildlife species by reducing protective cover. At the same time, thinning can have a positive effect on the production of herbaceous vegetation. The cover value of a stand depends on whether the trees are evenly or unevenly spaced. Thinning as a forestry practice is based on the hypothesis that trees in fully stocked stands follow an identical rate of self-thinning. The implication is that if the growing space is fully occupied, an increase in average stand diameter can only occur if there is a reduction in the tree density.

A loss of cover for wildlife is due not only to the reduction in the number of trees but also to the *plantation effect* (even distances between trees). In evenly spaced stands, the cover value is greatly reduced as compared to the same stand where trees are randomly spaced. Groups of trees in stands provide greater densities by creating microsites for cover. Thinning, if done with wildlife constraints in mind, can produce a stand arrangement that provides future benefits for deer, elk, turkey, squirrel, and other species.

3.3.11 FOREST REGULATION

The process of bringing a forest into balance to produce a sustained yield of wood volume is forest regulation accomplished by either area or volume control. In *area regulation,* a given size of area is cut each year or at other established time periods, but the volume between areas is variable. *Volume* refers to the amount of wood removed. *Volume regulation* is realized by removing a stipulated volume of wood each year or until there is an impact on other resources, regardless of the size of area cut. On the ground, application of the two control processes generally results in a combination of both to accomplish specific objectives. Regulation also involves rotation and cutting cycle.

3.3.12 ROTATION

Rotation is the time interval between one regeneration cut and the next on the same area of land to perpetuate an even-aged forest. Essentially, the rotation length defines the birth and death of a stand of trees in a successional process of going from seedlings to a mature stand that can be harvested. Rotation age varies by tree species but is a function of tree growth and management objectives. It is also guided by the age when a particular tree species reaches the culmination of *mean annual increment* (MAI). MAI is the age at which wood fiber is being added at the highest average annual rate the tree can achieve. It is not the greatest age or largest size a given tree species can attain. Given the following conditions,

$$\text{Area } (A) = 2,400 \text{ ac}$$

$$\text{Rotation } (R) = 120 \text{ yr}$$

the area to be cut each year is $A \div R = 2,400 \text{ ac} \div 120 \text{ yr} = 20 \text{ ac/yr}$ (Figure 3.9). One 20-ac stand is cut each year for 120 years and represents a fully regulated forest. The

Cut Year 1											
1	45	65	30	101	51	106	14	88	75	103	35
71	93	2	87	114	6	118	44	95	4	112	94
26	108	38	111	57	17	42	24	64	48	70	18
54	78	11	83	69	21	81	56	107	31	86	61
102	20	85	32	98	3	92	40	68	10	109	100
37	5	59	49	34	74	47	105	16	52	115	36
91	22	41	82	13	63	72	28	110	60	19	76
12	99	50	77	96	7	84	55	113	90	116	62
79	53	67	15	119	23	58	8	117	25	46	97
29	89	9	43	33	66	27	80	39	73	104	120

Cut Year 120

FIGURE 3.9 Regulation to produce even-aged stands. One 20-ac stand is cut each year over a period of 120 yr by clearcutting, shelterwood, or the seed tree regeneration methods. No stand of the same age class is adjacent to another stand of the same age class.

20 ac can be from one stand or from any combination of stands as long as the total equals 20 ac. For 20-ac stands, the result of the configuration is 120 stands distributed over the management unit (working circle, district, watershed, etc.) of 2,400 ac. In either case, the number of acres treated annually remains the same, but by cutting a smaller size stand, harvesting and transportation costs may increase to an unfavorable cost-benefit ratio.

3.3.13 CUTTING CYCLE

Cutting cycle refers to the interval of time between cuts in an *uneven-aged stand*. The idea is to cut the growth that has accumulated over the interval and then return 20 years later to the same stand to repeat the cycle. While even-aged management focuses on the length of rotation, uneven-aged management depends on the length of the cutting cycle. Given the previous example and using the following information,

$$\text{Area } (A) = 2,400 \text{ ac}$$

$$\text{Cutting cycle } (CC) = 20 \text{ yr}$$

$A \div CC = 2,400 \text{ ac} \div 20 \text{ yr} = 120 \text{ ac}$ in one stand to be entered every 20 years (Figure 3.10). All age classes are represented in the forest. The 20-yr entry can be made in any combination of stand size and number to equal 120 ac. For a rotation of 120 years and a cutting cycle of 20 years, six age classes are represented in the stand.

$$R \div CC = 120 \div 20 = 6$$

Assuming that each age class occupies equal areas across stands, then the forest is balanced with an intermixing of age classes. Stands created by the selection reproduction method will have great internal vertical but little horizontal diversity. However, this condition creates the potential for group selection, which in some cases can be small clearcuts.

The examples presented are hypothetical and probably do not occur in nature, but like the uninhibited growth model, they represent a base from which to compare the potential of a site if all conditions were present. Rotation, cutting cycle, stand size,

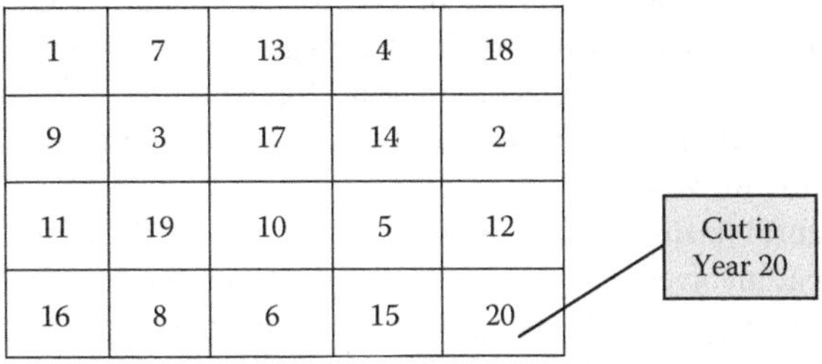

1	7	13	4	18	
9	3	17	14	2	
11	19	10	5	12	Cut in Year 20
16	8	6	15	20	

FIGURE 3.10 Regulation to produce uneven-aged stands. One 20-ac stand is cut by the selection regeneration method each year over a 20-yr period.

tree age distribution, horizontal and vertical diversity, along with local topographic features, desires of society, and ecosystem management issues would seem to present the practicing forester and wildlife biologist an almost impossible management situation because there are no "cookie-cutter" solutions. On the other hand, the potential of new combinations and doing business differently are beyond imagination while using adaptive resource management.

3.4 WILDLIFE MANAGEMENT

Wildlife was defined in Chapter 1. However, a definition of wildlife management is needed to complete the background to discuss the topics that follow. *Wildlife management* is the art and science of manipulating wild animal populations to meet specific objectives. The definition does not require a reason for managing wildlife for society typically decides which species are important. It is the job of management to *increase*, *decrease*, or *maintain* stable wildlife populations. The science part of wildlife management is based on sound biological and ecological principles for maintaining the health, sustainability, and productivity of animals in ecosystems.

Manipulating populations requires a basic understanding of the factors that affect the ability of an animal to survive and reproduce. The information provided in Chapter 4 adds to the general information on the factors defined in Conceptual Equation 1.1: hazards, diseases, predators, humans, species biology, and resources. These six factors are in the *centrum* of all wildlife species regardless of the land use type where they find habitat. Our emphasis, however, is on species in forest types, in particular how forest management practices affect their ability to survive, reproduce, and maintain a sustainable population.

3.4.1 HAZARDS

The modification of natural hazards such as rain, snow, and wind can only be accomplished on a small scale and with limited success, so in most cases their management is not a realistic consideration. Rain can be produced through cloud seeding, but the technique is expensive, and the results are too unpredictable to be useful at present. One aspect of producing rain, should it become practical in the future, is that large amounts of precipitation at one time may be detrimental to many wildlife species. For this reason, the effects of cloud seeding must be known before it becomes a practical tool. Snow movement can be controlled to some degree in open areas by installing drift fences or wind-rowing logging debris to provide open or lightly covered areas on the downwind side, making forage easier to obtain by deer and elk. Windbreaks such as have been developed for agricultural shelterbelts can also provide protection for wildlife.

Fire as a natural hazard is much more easily managed than rain, snow, or wind. Fire management involves planning the creation of fuel breaks in vegetation important for a particular species. In using fire, the identification of the areas to be protected is the first priority. The second job is to determine how large the fuel breaks must be and where they will be located to provide the most protection. The third task is to develop a schedule of maintenance, for if maintenance is not done the fuel break will lose its protective value over time. Fire, as a man-made hazard, results from arson or accidents from prescribed burns.

Before initiating a prescribed burn, the plant and animal species native to the area, along with any endangered and threatened species existing there, must be thoroughly documented. Special areas such as nest and roost trees, water holes, and browse concentrations are identified for protection by fire lines. Prescribed fires can be just as hazardous to wildlife as natural fires, but hazards can be reduced if wildlife requirements are considered in the planning and implementation of a prescribed burn.

In general, the effects of fire as a management technique can be summarized as follows (from Lyon et al. 1978; Wright and Bailey 1982; Kramp et al. 1983):

- Some browse species used by herbivores proliferate in early postfire succession, and their nutritional quality is high for several years after burning.
- In coniferous forests, fire stimulates plants that produce fruits and berries. A parallel increase in seed availability favors an increase in use of burned habitat by birds and small mammals.
- Wood-boring insects, some of which are important foods for woodpeckers and other insectivorous birds, may increase. Fires also may decrease populations of wildlife pest insects and parasites.
- Fire affects the scale and pattern of the vegetation mosaic through fire size, intensity, and frequency. These factors influence the relative abundance of plants and subsequent successional stages. The resulting patterns in turn affect herbivore populations.
- Carnivores dependent on certain herbivore populations benefit from a fire-created habitat mosaic because it increases the food base of their herbivores.

Man-made hazards are primarily roads and fences, which are marginally easier to control than natural hazards. Road hazards can be reduced in the planning process by avoiding alignment through areas critical for survival and reproduction of wildlife species. In addition, roads can be fenced and underpasses built to provide for free movement where a road separates two use areas or crosses a migratory route. Road hazards can be further reduced on high-speed highways by selecting unpalatable cover plants for roadside and median plantings. The opposite holds true for unpaved forest or other infrequently used roads. Also, roadkills along major highways can be reduced by placing 90° light reflectors about 10 m (33 ft) apart along both sides and in the median.

Fences, depending on the type, are obstacles to free movement of wild animals. Barbed wire is especially hazardous for juvenile antelope, deer, and elk. This problem can be reduced by using smooth wire on the top and bottom strands of the fence. High fences of wood or hog wire can be installed to direct animals to desirable locations or away from undesirable locations.

3.4.2 DISEASES

The control of wildlife diseases is a subject that has not been given the attention devoted to other decimating factors. In part, this lack of attention indicates a widely held belief that wild animals cannot be treated for diseases, so why waste time and effort in trying. Disease is a natural way of controlling animal populations by eliminating unhealthy animals. Disease-weakened animals are easy prey for predators that otherwise would try to kill healthy individuals. Leopold (1933) believed that the effect of disease on wildlife was underestimated, and this is probably the case even today. Diseases that clearly kill wild animals are difficult to identify except when there is a large die-off or symptoms are discovered during routine checking of animals killed in a hunting season.

Not all diseases and parasites kill animals directly. Some can cause side effects that may be objectionable to hunters. For example, botfly larvae parasitize the eastern gray squirrel, causing nonlethal wounds that hunters find so unpleasant they discard the squirrel carcass (Jacobson et al. 1979). The manager's remedy for this problem is to postpone the squirrel hunting season from September to October, after the botfly larvae have left their host and the objectionable wounds are healed. Cottontail rabbits not only suffer from the same botfly problem but also carry ticks, which cause tularemia in rabbits and humans (Yeatter and Thompson 1952). Again, managers can reduce the incidence of the disease in humans by having a hunting season during cold months when ticks have left the host and shortly after all infected rabbits have died.

At one time, there was great concern about meningeal worm infections of white-tailed deer. Research results later showed that the parasite had little effect on white-tailed deer, but it is known to cause death in other ungulates, such as mule deer, moose, reindeer, and elk (Anderson 1972; Anderson and Priestwood 1981; Davidson et al. 1981). This particular parasite is transmitted by snails as the intermediate hosts. The solution to the problem is to find a way to control the infected snails.

In the case of highly prized trophy species, managers have found it desirable to capture animals and treat them with drugs to ward off disease. This was done in the case of bighorn sheep in Colorado (Schmidt et al. 1979). This technique is expensive and time consuming, and the results are difficult to evaluate, but benefits to management through increased survival may make it worthwhile. However, it is hoped that through research, various diseases might be associated with habitat quality and thereafter techniques developed to prevent many diseases through habitat manipulation. In this regard, manipulating habitat with the objective of increasing a population may cause concentrations of animals in which the transmission of disease is enhanced. As can be seen, the control of diseases and parasites in wild populations is at best a difficult, but mostly an impossible, task.

Herbicides and pesticides (insecticides, rodenticides, etc.) are included with diseases because, like diseases, they affect the physiology of animals. But, unlike diseases, they can be controlled because they are manufactured and sold by humans. One of the problems involved with these two groups of compounds is the attitude by some users that if 1 kg per hectare will do the job, then a higher concentration will do it better. Herbicides and pesticides are now licensed for use by government agencies (Environmental Protection Agency in the United States) after a period of testing and documentation of the toxic effects. When using herbicides and pesticides, it is incumbent on the user to follow the prescribed precautions listed on the label. The trend now is to develop herbicides and pesticides that biodegrade in a short time and to develop natural (biological) techniques to control the target organism or vegetation.

3.4.3 HUMANS

The management of humans as a directly acting factor of Conceptual Equation 1.1 centers on education and legal regulation. Education provides information to understand the role and utility of animals in the functioning of ecosystems. Education is also involved with regulations aimed at protecting resources necessary for the survival of both humans and wild animals. Humans are ecological dominants, so they have to be controlled because a few are given to exploiting the wildlife resource. With modern tools and weapons, humans can do great damage in a short time. Conversely, others choose to protect everything. In such circumstances—with vocal partisans at both extremes—conflict inevitably arises. Managing humans, as most managers soon realize, is generally more difficult than managing the wildlife resource. In the past, natural resource managers have not received the training required to deal with the human dimension. But, in recent years managers have placed greater emphasis on understanding human behavior and the values of wildlife as perceived by the citizenry (Kellert 1980).

Colleges and universities are also doing a better job in educating resource managers in psychology and operations research. Further, government agencies are now providing training in conflict resolution to field managers and enforcement staff. As a result, managers are more aware of the human dimension in affecting changes in management strategies, which in turn will affect wildlife populations, habitat, and human experiences. A large number of resource-oriented organizations influence their members and others through educational programs and involvement in the formulation of conservation and regulatory laws. They include professional societies such as the Society of American Foresters, the Wildlife Society, and the Society for Range Management; conservation organizations such as the National Wildlife Federation, National Audubon Society, Ducks Unlimited, Wild Turkey Federation, Rocky Mountain Elk Foundation, and Sierra Club. In many cases, the members and officers influence resource management agencies in setting goals and objectives and in monitoring situations and conflicts that they consider important to their interests.

Government management agencies depend on conservation organizations to support their objectives because it is often necessary to enlist their help in securing funds for research and management purposes. Indirectly, this is a form of people management since the agencies must convince the organizations that their goals are sound and will benefit the public. The manager or biologist demonstrating a good understanding of wildlife and how to deal with the people involved with wildlife, either directly or indirectly, will be the successful manager.

3.4.4 PREDATORS

Predator control as a wildlife management technique is largely oriented toward those animals that prey on game species. However, modern ecological thinking that focuses on the food chains and webs involved in nutrient cycling and energy flow in ecosystems strongly opposes the killing of every predator. As a consequence, predator control has awakened public concern and criticism, making it desirable to conduct hearings before any large-scale control program is undertaken. The history of wildlife management contains many examples of failed and successful predator control programs. This past experience has led to the development of some general principles useful to managers in the decision-making process.

- As a management tool, predator control is generally not feasible over a wide area for both biological and economical reasons. From an economical point, the cost-benefit ratio must favor predator control if it is to be a workable management tool.
- Predator control is best used only to solve a local and temporary problem, such as unusually heavy predation in a refuge, fish hatchery, or other area where the prey species is found in unusually high numbers (poultry farms).
- In many cases, predation results from the behavior of an individual animal, not the species as a group.
- Intensive predator control programs do not result in an increase in game commensurate with the control effort and expense in most wildlife management areas.
- A predator control program often inadvertently increases the populations of other species, such as small rodents.
- Predator control must on occasion be initiated for health reasons, such as control of rabies by reducing populations of foxes, skunks, or raccoons.

One example of a study to determine the cost-benefit ratio for predator control was done in Arizona in pronghorn habitat. The conclusions were that high fawn survival was correlated with the removal of coyotes by helicopter gunning in the spring prior to fawning. The cost of control compared to the value of the increased number of pronghorns for hunting was determined to be favorable (Smith et al. 1986). Methods to control predators are varied, but programs usually involve one of three methods:

- Professional hunters and trappers
- Bounty systems open to the general public
- Poisons

There are advantages and disadvantages of all three methods. Poisons are not selective to a species, bounty systems are open to fraud, and professional hunters and trappers are expensive. Most game and fish departments now use their own employees to control predators after intensive review and documentation of the environmental effects.

3.4.5 BIOLOGY

Biology is an innate genetic characteristic in the centrum. While centered in the individual animal, biology influences the development of populations. Management of species includes management of populations by stocking, refuges, and harvesting. One of the first techniques used in wildlife management in the early 1930s and 1940s was stocking of game animals. Stocking includes both the introduction and the reintroduction of a species. As with predator control, there have been many successes in stocking animals, but wildlife literature also contains examples of unsuccessful attempts. Stocking is not a substitute for habitat improvement, and the need for stocking must be examined carefully before initiating this expensive management technique. Some of the reasons for stocking are

- To populate an unoccupied habitat or reintroduce an extirpated species. *Extirpated* means driven out or eliminated in a specific area. Such stocking programs using animals trapped in other areas have been successful with some species such as deer, turkey, elk, and bighorn sheep. The unoccupied habitat must be suitable for the introduced or reintroduced species.
- To bolster a declining population. Stocking has rarely been a successful means of accomplishing this objective as the deterioration of habitat quality usually accounts for the decline in native populations.
- To increase the number of animals available to hunters. The cost of this method in areas other than a hunting preserve makes it impractical. However, some government agencies have found that "put-and-take" fishing can be cost-effective.
- To introduce a new species, usually an exotic. An *exotic* is any animal not native to the area where it is being introduced. Several exotics have been successfully introduced into North America, including the ringneck pheasant, chukar partridge, Hungarian partridge, rainbow trout, and several African species in New Mexico.

The last objective differs from the first three because of the greater biological risk involved in introducing exotics, which may compete with native species. Many exotics, such as rabbits in Australia, eastern grey squirrel in England, English sparrows in North America, and deer in New Zealand, have caused serious management problems; in these three cases, the introduced species had no natural predators, so populations are now out of control, with quite disastrous results—particularly in the case of rabbits and deer. Before introducing exotics, managers should ask these questions:

- Is the niche occupied by a native species?
- Will the exotic become a pest?
- Will it bring in disease?
- Does it have recreational, cultural, or aesthetic value?
- Is there habitat of suitable quantity and quality to meet the life requirements of the species?

An outgrowth of the introduction of exotics has been game ranching, by which animals are raised for food. This method has been tried in several African countries because of the need to provide protein for undernourished people. One of the advantages of using wild animals in such a program is their ability to reproduce and survive much better than domestic cattle. When a decision has been made to introduce an animal into unoccupied habitat, careful thought and attention must be given to the source of animals for stocking. This should include information about where the animals will be obtained. Will they come from game farms or trapped wild animals? In most cases, game farm animals are unable to take care of themselves when released.

Trapping wild animals is expensive, but the animals are better able to survive and reproduce. In the case of birds, there is also the possibility of using foster parents or double clutching. For example, Texas bobwhite has been used to hatch masked bobwhite eggs. In *double clutching*, eggs are taken from the nests of wild birds to get them to lay another egg. The egg that is removed is hatched artificially and the chick raised without human contact to be returned to the wild. This technique has been used successfully with endangered species, such as the peregrine falcon.

The notion of establishing *refuges* was originally based on the premise that a protected population would increase and overflow into adjacent areas. Refuges are used as a management area to protect one or several species, while a *hunting preserve* is used to raise animals for sport hunting. The refuge system has worked well for highly mobile species and for species needing special protection. Refuges are generally established for the following reasons:

- Nesting, feeding, or loafing areas
- Escape areas
- Protection of an endangered species
- Research purposes

Refuges established in the waterfowl flyways have been the most successful. Income from the sale of waterfowl stamps is used to acquire wetland to establish waterfowl refuges.

Harvesting is the principle management tool in the control of game animals, and traditionally it has been controlled by the regulation of hunting seasons designed to kill excess populations created by management. While harvesting of animals is often a controversial topic, it is still done in every state in the United States, in every Canadian province, and in most other countries. Some of the objectives of regulating hunting seasons are

- To structure the harvest in such a way that the maximum number of animals is provided to the maximum number of hunters.
- To regulate the number of animals killed to ensure the continued productivity of the hunted species. The regulation of numbers killed in a hunting season involves controlling:
  a. The timing and duration of the hunting season
  b. Daily and seasonal limits
  c. Licenses or tags sold, rationed, or auctioned to hunters
  d. Types of hunting equipment permitted

All the mechanisms indicated have the effect of reducing hunting pressure by applying restrictions at different degrees of intensity. The wildlife manager therefore has a variety of tools to limit the number of animals killed in a hunting season. In addition, some of the controls, such as types of equipment and restricting the number

TABLE 3.4
Approximate Percentages of a Population that Can Be Removed by Hunting

Species	Percentage
Wild turkey	25–35
Deer	5–10 (bucks-only season)
	20–30 (to hold population stationary, any choice season)
	30–40 (to reduce population, any choice season)
Rabbit	60–70
Squirrel	25–50
Bear	5–15
Quail	30–40

of hunters by licenses or tags, also contribute to hunter safety. When determining hunting regulations, the following general principles of harvesting have proved useful in guiding the control of hunting:

- A given amount of hunting pressure takes a larger proportion (%) of a population when it is high than when the same population is low. For example, a 10% harvest from 1,000 animals is 100, but a 10% harvest from a population of 100,000 is 10,000. The 100,000 animals may be able to withstand the 10% harvest, but the 1,000 animals may not.
- Any game species characterized by a high rate of increase can normally sustain a high rate of kill (Table 3.4).
- In some instances, a high kill rate stimulates production in a given game species.
- The law of diminishing returns often sets the harvest for some species. As more effort is needed to hunt to find an animal, the number of days hunted and the number of hunters declines.

Hunting has come under attack by many environmental, ecological, and animal rights groups, primarily trying to stop the killing of big game animals by legal action or by active protest. The question is: What would happen if all hunting were stopped? Initially, there may be an increase in the density of game animals, but there also would be a loss of millions of dollars in license fees, which now contribute to maintaining habitat of many nongame wildlife species. Eventually, without hunting there probably would be some steady state reached at which wildlife populations will reach dynamic equilibrium between animal density and habitat conditions.

Humans, as a result of an ancient need to hunt for food and fur persisting over thousands of years, may have developed a psychological need or at least a predisposition to hunt. Although most hunters in developed countries do not need to hunt for food, the compulsion to do so remains ingrained in the psyche of many. The hunting experience also seems to provide a sense of becoming a part of nature in which the role of predators is acted out to achieve some psychological benefit that is difficult to explain (Copp 1975).

Before a hunting season can be set, some accounting must be made of the yearly population cycle of the animal hunted so that the numbers to be harvested can be estimated. The life history of the species provides considerable background information, but it is necessary to condense and reformat the information to develop a useful management tool. This synthesis of information yields a *life equation*. Life history deals with events over the entire life of an animal. A life equation accounts for changes in a population as a result of yearly life events.

The best way to understand a life equation is to construct an example, but first several points must be made. A life equation should be flexible and generally applicable to a wide range of habitats, all of which on the average are capable of supporting a population. These populations should, in the absence of hunting, continue to increase in density until hazards, disease, predators, and resource shortages bring the population into equilibrium with its environment.

Second, these generalized life equations must be modified by the manager to reflect the actual conditions found in individual management units. Third, life equations should represent average conditions, especially for weather, which exerts a tremendous short-time effect on total populations of some animals, such as quail and rabbits. Finally, life equations do not generally take into account cyclical phenomena exhibited by some species. They do not take account of random ecological changes by floods, timber harvesting, or vegetation pattern changes even though in long-range planning such factors are important and must be considered. This difference arises because the life equation is recomputed on an annual basis to set hunting seasons, while the time frame of long-term planning extends over a number of years.

A review of a hypothetical life equation example based on the wild turkey using generalized and not actual data can be found in Table 3.5. The wild turkey life

TABLE 3.5
Hypothetical Life Equation for Wild Turkey

Time	Activity	Young	Adult
Spring	1. Desirable base population is 2 males and 8 females.	—	10
	2. Clutches average 15 eggs, 8 × 15 = 120.	—	—
	3. 80% hatch.	96	—
	4. Sixteen die for various reasons.	80	—
Summer	5. Thirty die from hazards.	50	—
Fall	6. Fifty young become adults.	—	60
	7. Ten are lost to predators.	—	50
	Assuming a 50:50 sex ratio, 25 males and 25 females remain.	—	—
	8. Thirty birds will die over the winter, so the legal harvest can be 30.	—	30
Winter	9. Remaining birds after harvest.	—	20
	Ten birds are a safety factor.	—	—
Spring	10. Two males mate with 8 females.	—	10

equation is typical of the annual cycles of virtually every animal population. If 2 males mate with 8 females and the females lay an average of 15 eggs, ideally up to 120 eggs could be hatched. But for several reasons, only 96 (80%) hatch, and of these, 16 chicks die for various reasons. Another 30 chicks are lost to hazards. Of the 120 eggs laid, 50 chicks live to become adults in the fall, which, added to the original 10 birds, yields a total of 60 in the population. Of these, 10 birds are killed by poaching or predators. Assuming a 50:50 sex ratio, 25 male and 25 female adult turkeys remain.

Experience has shown that a spring population of 2 males to 8 females will maintain the base population as a sustainable population, and that over the winter there will be a loss of about 30 more turkeys from the earlier population of 50 birds. These 30 turkeys at the time of the fall hunting season are surplus to the management unit's capacity, so they can be killed by hunters, leaving 10 birds for spring breeding and 10 as a safety factor. This example does not take into account a common practice of a spring gobbler season. Even though the life equation is hypothetical and simple, it does illustrate the way in which a population is analyzed to determine the surplus that management has developed for the benefit of hunters. Actually, given the uncertainties of environmental factors, getting to 2 males and 8 females in the spring is a major problem, but that does not lessen the usefulness of the technique.

In an unhunted population, the difference between the uninhibited growth model and the inhibited growth model is the number of animals that are lost due to diseases, hazards, and other factors. When a population has reached ecological carrying capacity, there is still a seasonal fluctuation in animal density. In the spring, population density is low because of mortality, but in the fall it is high as a result of natality. The mortality occurring from fall to spring is a huntable surplus.

The idea is that a surplus can be removed; otherwise, it would be lost to winter mortality. The surplus of game animals killed each year is referred to as *yield*. *Sustained yield* is the number of animals that can be removed from a population over a period of years, leaving a breeding population sufficient to maintain not only the equilibrium number of animals but also a harvestable surplus. This conclusion assumes that the number of animals killed by hunters is compensatory. The concept is that the total effect of mortality does not vary, but different types of mortality, such as those caused by disease, predation, or resource shortages, can each vary, and one will compensate for the other. Hunting is hypothesized to replace the mortality caused by other decimating factors and seems to hold true for species with high natality, such as small game, but may not hold for big game species.

In the inhibited growth model, the maximum rate of production occurs at the inflection point, which is about 50% of the carrying capacity of the habitat for the species. It is the point at which growth is increasing but at a decreasing rate. If the population is reduced each year to this level, then the recovery rate will be at a maximum, which is called the *maximum sustained yield* (MSY). Several problems make it difficult to apply MSY in a field situation. One is that K is almost never known, and a statistically valid estimate for P is difficult to obtain even under the best of conditions. The MSY has also been criticized as not directed at maintaining ecosystems (Holt and Talbott 1978). Further deficiencies of MSY include

- Focusing only on the population of a single species
- Assuming dependence on the ability of a species to adjust that may not exist
- Emphasizing yield without regard to the effort needed to produce the yield
- Not taking into account fluctuating yearly conditions that produce variable vegetation growth
- Not including an insurance factor to ensure that a viable population will always exist

To compensate for the shortcomings of MSY, Holt and Talbott (1978) offered four suggestions:

- Ecosystem maintenance must be optimized so that consumptive (hunting, fishing) and nonconsumptive (photographing, viewing) values can be maximized on a long-term basis, present and future options are ensured, and the risk of irreversible change or long-term adverse effects resulting from human use is minimized.
- Management decisions must include a safety factor to allow for the fact that human knowledge is limited and human institutions are imperfect.
- Measures to conserve all wild living resources should be formulated and applied to avoid wasteful use of other resources.
- Survey or monitoring, analysis, and assessment need to precede planned use and accompany actual use of wild living resources.

The reality is that public desires and opinions may control the annual yield regardless of the intent of science. Public desires and opinions translate into lobbying of elected officials. Mixing of biology and politics is appropriately called *biopolitics* and is of major importance in the decision making of all wildlife managers.

3.4.6 RESOURCES

The abundance of wildlife in a *terrestrial* habitat depends heavily on the adequacy of food, cover, water, and space and the desires of people. Habitat management for a given species involves the manipulation of vegetation as the most important factor contributing to maintaining a viable population. Even carnivores are indirectly dependent on plants for energy since most of the animals they eat are themselves herbivores. Each forest successional stage provides different types of food and cover that satisfy the minimum requirements of some species, but not all, and yet provides optimum conditions for others. Thus, several species may have all their requirements met in a single stage, while some require one stage for food and another for cover. Because seral stages progress over time to climax (Axiom 2), the vegetation cycle of food and cover for different wildlife species occurs on the same tract of land. Vegetation is controlled by nature but can be influenced by humans.

In the absence of human interference, vegetation will naturally develop in a balanced way that will always contain habitat components for a number of species. A balanced system is what one would expect in undisturbed wilderness areas, but in reality vegetation is always changing even though the rate of change may be difficult to detect. Under human influence, management can be directed to maintain vegetation for a single species or a group of species with overlapping habitat components. Using this approach, vegetation can be kept in or directed to a seral stage that produces the combinations of food and cover required by one or more wildlife species. This is done by directly interfering with succession to return stages to a lower level of secondary succession.

Another alternative, if the environmental factors are suitable, is to convert the area to a completely different vegetation type, such as forest to grassland or hardwoods to conifers. Control of wildlife populations is directly the responsibility of game and fish agencies and is usually accomplished by regulating hunting seasons and bag limits. However, wildlife populations can be controlled indirectly through habitat manipulation, an activity that is the responsibility of state, federal, or private land management agencies. Direct techniques used to modify succession in forested areas are

- Prescribed fire
- Controlled grazing
- Timber harvesting
- Using herbicides
- Seeding and planting

In addition to the techniques employed by humans, nature has from time immemorial controlled succession by wildfires, such as those in Yellowstone National Park in 1988, and disease or insects, such as the bark beetle outbreak on the Coconino and Kaibab National Forests in 2007. While humans may manipulate vegetation for scheduled brief periods of time, natural forces are always at work to cause changes in vegetation structure. Within the large land areas classified as a forest, a great many smaller communities and ecosystems can be identified. Among them are the *aquatic resources* of marshes, riparian zones, and habitats of emergent and submergent vegetation, all of which add to the total diversity and wildlife productivity of the forest type.

The management and protection of watersheds, as the source of water for streams and lakes, have always been important functions of forest management. Watersheds control water quality. When maintained in the proper condition, they prevent soil erosion and serve as filters for the streams within the watershed boundary. Foresters and wildlife biologists are concerned with the effects of various watershed uses and treatments on the natural character of the stream environment providing fish habitat. In cold streams running through forested areas, trout and salmon are considered the most important game fish, so the discussion that follows focuses on these two soft-rayed groups.

Most *spiny-rayed fish* are tolerant of a wide range of water conditions, but soft-rayed trout and salmon are intolerant. Trout and salmon are most often found in streams characterized by the following conditions favorable to their survival and ability to reproduce (Forbs 1961):

- A good stream is a relatively fast-moving body of clear water averaging about 14°C (57°F) in the summer months, with deep pools at intervals along its length.
- The stream must have frequent spawning areas of coarse, clean gravel in moderately fast water.

- The stream banks must be well forested and well vegetated for food production and temperature control. For streams 1.5–6.0 m (5–20 ft) in width, a cover of 50–75% of tall trees and shrubs is a necessity. Cover along wider and deeper streams is not as important since water depth itself furnishes cover.
- The stream should have about 50% pools and 50% riffles, well spaced. The riffles are food-producing areas, while the pools provide hiding spots.
- The stream should not be subject to flooding sufficient to cause lateral course changes each year. Normal spring flooding may not be a cause for concern.

The factors initiated by humans that directly affect trout streams are mainly logging, grazing, and in some cases high recreational use. Where forests are removed in large clearcuts, the volume of water is greatly increased. In addition, spring runoffs are regularly larger, while summer water flows are typically lower following a heavy timber harvest. The key to protecting forest streams is to harvest small areas over longer time intervals to balance water flows. However, if larger permanent water flows are needed and would benefit trout waters, then heavier harvesting, if properly done, may improve water temperature and depth.

Livestock, deer, elk, and moose tend to concentrate much of their use close to or in riparian zones. Overuse of streamside vegetation by grazing animals contributes to the destruction of stream banks and streamside vegetation, thereby increasing erosion. Fencing has been used to restrict livestock access to streams, but deer and elk can easily jump a fence. Successful management under these circumstances requires a balanced multiple-use approach that includes limiting livestock and wildlife numbers together with routine watershed maintenance.

Recreational campground use is often the cause of erosion and stream bank deterioration. In addition, water quality in the absence of sanitation facilities is adversely affected. Considerable progress has been made in monitoring water quality as a result of the National Environmental Policy Act. However, an increasing human population, resulting in more use of outdoor areas, necessitates constant monitoring of water quality in heavily used areas. In some situations, erosion can be controlled by building structures to stabilize the banks or by plantings aimed at the same goal. Other control measures include limits on campground use.

3.5   INTEGRATING FACTORS

Considerable information has accumulated on the effects of timber harvesting on various species of wildlife in different geographical areas. Intuitively, it has long been understood that timber harvesting can have positive, negative, or neutral effects on wildlife habitat depending on the life requirements of the species inhabiting the area. The effects of timber harvesting on wildlife are best judged by what actually happens to the habitat of an animal as compared to its habitat requirements. The main concern is not which harvest method will be used or how many trees will be removed, but **how many trees will remain**?

Both the professional wildlife literature and the news media commonly invoke the words "effect on wildlife" when discussing forest management practices such as cutting, burning, and spraying. In general, this association without naming the specific species affected commonly results in a misunderstanding of the problem by other disciplines and the general public.

3.5.1   A MATTER OF SCALE

Populations of animals of narrow ecological amplitude and specific habitat requirements (amphibians, reptiles, small birds, and mammals) can be adversely affected at the time of a timber harvest even if the cut is limited to a small area or to a single tree. The tassel-eared squirrel, a moderately mobile animal, builds a nest in a single tree, so the removal of that tree is a direct loss of a habitat component. A loss of a nest tree does not imply that an animal using the tree is destroyed, but it must use energy in selecting a new location for building a nest. This energy could otherwise be used for body maintenance and reproduction.

Highly mobile animals with a wide ecological amplitude (large birds, mule deer, and elk) are least affected by a small cut or the removal of a single tree in one stand. As the size of cut area increases, the food-to-cover ratio changes, thereby providing a positive benefit, but only to the point at which the amount of cover decreases to the minimum required to maintain a viable population. The gain in forage for deer and elk from the removal of one tree or a small group of trees is not significant unless similar cuts are repeated many times in the management unit.

3.5.2   TIME

The successive seral stages occurring during a rotation progress through time from either primary or secondary succession to climax. The result is a change in wildlife food and cover accompanied by a change in wildlife species on the same tract of land. A resource manager thoroughly familiar with the time requirements for development of the seral stages of a particular vegetation type can simulate the location of food and cover and therefore the presence of wildlife species for any future time period. Because humans tend to think in terms of the "here and now," difficulties arise in visualizing the concept that timber harvesting is a process that resets the renewal cycle for vegetation and animals back to another ecological period. Theoretically, present forest conditions can be duplicated in the future if the correct silvicultural prescriptions are applied when harvesting and regenerating a stand of trees.

3.5.3   NATURAL ROTATION

Removing old or mature trees in a stand every 20 yr does not provide habitat for animals needing "wolf trees" (trees with large spreading crowns), snags, or old forest conditions. Therefore, several stands considered for harvesting and some groups with old trees in other stands can be reserved for continuous use by wildlife. These stands and individual trees can be allowed to go through a *natural rotation* of growth and decay, which may be as long as 300–400 yr or longer depending on tree species. Even in a natural rotation there is birth and death of trees, and when the old forest conditions fade, there is room for regeneration to restart the process.

By preserving old forest conditions and decadent trees in other stands, conditions can be perpetuated for a time to meet the needs of wildlife species depending on older, mature trees. The trees saved ultimately become the wolf trees and snags used by many nongame bird species and small mammals. The same principle can be applied to even-aged management systems in that some stands and trees are reserved to go through their normal life cycle, which is beyond rotation age.

Because factors such as topography, site quality, and climate can lead to deviations from what is possible, the examples presented (Figures 3.9 and 3.10) are ideal but may never occur. However, they are the consequences of silvicultural and forest regulation that provide the foundation for practical application of forest science. The manager's job is to consider all these factors in designing a system to perpetuate the plant species being manipulated to meet the needs of both humans and native wildlife species for an indefinite period of time involving hundreds of years into the future.

3.5.4 STAND SIZE AND DISTRIBUTION

Conforming to a strict stand size and number to meet area requirements does not provide the diversity that is needed to maintain habitat for a large number of wildlife species. The question revolves around maintaining

- A large number of small stands,
- A small number of large stands, or
- A variety of stand sizes from large to small.

Another question arises from the three alternatives: What stand sizes would include the home range of local wildlife species (Figure 3.11)? Landscape and wildlife diversity is greatest with the third option, and the problem becomes one of how many large and how many small stands (or openings) to create or perpetuate.

3.5.5 LOG-NORMAL DISTRIBUTION

One alternative is to distribute stands sizes to follow a *log-normal curve* because it occurs frequently in nature; that is, there are many more small species of animals than large ones, more small fires than large fires, more small trees than large trees, and so on. The idea is to mimic sizes and distributions of natural processes and disturbances that occur in forested areas (Harris 1984). Disturbances are not planned, and the reaction to major natural events from large fires to small "blowdowns" is to manage such instances by adaptive resource management (Walters 1986). However, managing outside of natural disturbances is through a defined planning process using the best information available to set a direction based on a time interval. This direction is often regulated by federal laws with established guidelines as was the case in the NFMA. The legal requirements for habitat distribution in national forests in the United States are contained in the NFMA (Mealey et al. 1982).

To summarize, the NFMA implicitly establishes the legal requirement for habitat dispersion by setting maximum size limits for areas to be regeneration harvested in one operation (Sec. 6(g)(3) (F)(iv)) and by requiring that such cuts be carried out in a

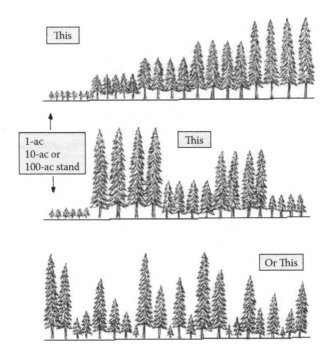

FIGURE 3.11 Which tree and stand size and distribution is best for wildlife?

manner consistent with the protection of soil, watershed, fish, wildlife, recreation and aesthetic resources, and the regeneration of the timber resource. Maximum size limits on cuts require that some portions of some harvestable stands remain uncut. This imposes some degree of scattering of harvest blocks among uncut areas. Compatibility of such cuts with the protection of wildlife resources demands a certain amount of edge and retention of cover which are necessary for wildlife. Effective edge and cover in timber harvest areas result from adequately scattering cuts through uncut areas.

3.5.6 Nonadjacency Constraint

The distribution problem can be approached by identifying the stands to be cut based on the constraint that adjacent stands must be of different age and size class. A *nonadjacency constraint* requires that stand locations are to be identified and delineated on aerial photographs and maps that represent the real world. There is some level of probability that two adjacent stands of the same age or size class might result, thereby creating a larger habitat unit in a specific age class; however, in some cases the larger unit may be desirable depending on the indigenous wildlife species. The shape of the boundary of adjacent stands imposes constraints in the planning and creation of a mosaic of stands. Squares are the easiest to consider since mathematically the minimum number of age/size classes needed to ensure that no two are adjacent is two. The same mathematical theory requires only two colors to create a checkerboard, but checkerboard patterns can lead to a forest without a distribution of stand sizes.

3.5.7 THE LANDSCAPE FACTOR

The log-normal curve is intellectually appealing because it relates to events and things that occur naturally; however, its actual application on the ground may be difficult. The difficulty arises from the fact that landscapes in mountainous country or rolling hillsides have microareas that are variable in soils, moisture, and plant species, and this variability will affect stand size. In reality, there is no predetermined size and distribution scheme that will work for all situations. As a result, *landscapes should dictate the size and number of stands because the "lay of the land" will influence management direction and here-and-now decisions.*

Large clearcuts do not have to be in squares or circles but can be long, narrow strips that blend into the landscape along a contour or topographic feature (Figure 3.12). An elongated clearcut could benefit many animals if there were different age classes along both sides of the clearcut. Over time, different stands can be created along the sides of the elongated area. Harvests can be arranged by visual adjustments on maps or aerial photographs so that no two stands of the same age or size class are adjacent, but the final distribution may require one or two rotations to achieve. By this process, the stands can be rearranged to create considerable contrast between stands.

Other more restrictive constraints can be adopted, such as requiring arbitrary differences between adjacent stands such as age classes, average stand diameter,

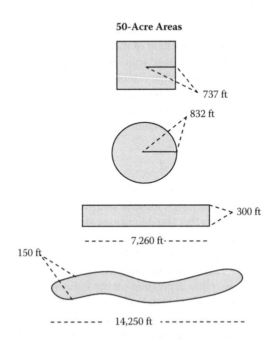

FIGURE 3.12 Comparison of a block, circle, regular rectangle, and an elongated rectangle clearcut that follows the natural terrain. The rectangles reduce the distance to cover from the center of the cut width.

basal area, or vegetation height. When restrictions are applied, the arrangement of the stands on the ground becomes more complicated and difficult to realize in one rotation without violating the restriction (Gross and Dykstra 1989). Restrictions can reduce outputs and value for other resources in addition to wildlife habitat. For a featured species or group of species selected for management, an important aspect of the effects of timber harvesting on wildlife habitat and populations is: *How much food and cover remain following a timber harvest*?

It does not matter which harvest method is used to regenerate a stand as long as the scale of the factors involved (time, management unit size, size and intensity of harvest, etc.) meet the critical food, cover, and space needs of the species to be managed. Since different vegetation types are found in diverse soil, rainfall, and temperature conditions, it is not practical here to prescribe silvicultural treatments appropriate for all forest areas and all wildlife species. However, some general guidelines that can be followed in selecting a suitable harvest method to benefit wildlife in any region can be stated in terms of scale:

- The size of the harvest area in the management unit
- The habitat requirements of the wildlife species in the management unit
- Successional patterns of forest stands during a rotation
- Size and intensity of a harvest, including the physical structure of stands
- Size and juxtaposition of openings remaining after the timber harvest

The joint practice of forestry and wildlife management must concentrate on maintaining ecosystems at the *landscape* or *macrolevel*. This approach necessarily calls for adjustments at the micro- or stand level to encourage diversity of plant and animal species within the landscape to ensure the functioning of the whole system. Such a view suggests that any given stand cannot be managed in isolation but must include an assessment of the impacts of management on adjacent stands. Along the same lines, one animal species, with some exceptions (i.e., endangered and threatened) should not be managed in isolation but rather in connection with the other species inhabiting the ecosystem.

Wildlife managers must adopt a multispecies philosophy, taking into account ground squirrels and chipmunks as well as small passerines when considering habitat improvement projects for game animals such as deer, elk, and turkey. Fortunately, this orientation is becoming more common, and in some state and federal agencies genuine progress in multispecies ecosystem consideration is now evident.

3.5.8 TRADITION AND OTHER CONSIDERATIONS

While a conflict between the harvesting of trees and the provision of habitat for forest wildlife species often occurs, there are also opportunities to improve habitat for many species. Timber harvesting to produce wood products can have either negative or positive effects on wildlife inhabiting the cut area; it all depends on the scale and intensity of harvest (Axioms 2 and 4) and in what forest cover type. However, the forest biologist and silviculturist working together can usually arrive at acceptable compromises to meet the objective for both forest products: wood and wildlife.

As a result of advanced technology in satellite imagery and aerial photography, it is easy to ground truth and document stand location and boundaries for current and future reference. Historical records of the condition of a stand and the decisions made on silvicultural treatments will help determine what is needed to set direction to meet planned goals and how a stand fits into a harvest schedule to meet these goals. Techniques to collect and store accurate inventory information at the forest landscape level are available, and a landscape-level inventory should be a high priority for preparing any forest management plan for timber and wildlife integration.

In a timber harvest, there are many ways to mitigate conditions brought about by harvesting activities. Mitigation is often by direct intervention or through coordination measures that will reduce the impacts of removing trees. The habitat management practices and coordination measures are discussed elsewhere (Section 3.6.1). Coordination measures are the basic practices to review when looking for habitat improvement opportunities in timber sales. Understanding these measures, and the factors involved in the scale of timber harvesting, will strengthen the forest biologist's position in making recommendations for mitigating negative effects or for habitat improvement when reviewing timber management plans.

Many combinations of cutting cycles and rotations in even-aged and uneven-aged forest management as well as many combinations of silvicultural regeneration systems can be used together. It is no longer necessary to be bound by the tradition of removing every tree from a clearcut or necessary to remove all down and dead material, as well as all dead standing trees, when using the other harvest systems. Wildlife managers are beginning to perceive that creating edges need not always be the sole guiding principle of management; some exceptions to the rule provide a better balance between vegetation heterogeneity and homogeneity because both relate to sizes and shapes of stands. Also, it is understood that every acre of land will not contain space for all wildlife species.

Livestock and wildlife conflicts are the direct result of both groups of animals using the same forest type for food and cover. Ranchers lease rangelands from state and federal governments for livestock grazing. These same lands, however, are also used by native wildlife species. Some ranchers therefore claim that deer and elk are using forage that is reserved for livestock. State wildlife agencies hold the opposite view.

Today, where public lands are managed under a multiple-use concept, livestock grazing is permitted but is more or less controlled. However, in many areas the livestock-wildlife conflict is as intense as the timber-wildlife conflict. An example of the type of problem that can occur is the 1989 incident of an Arizona rancher killing elk because he claimed they were using forage that belonged to his cattle. It is common to broadly group forests and rangelands for statistical reporting because the same resource agencies often manage both. A forest was defined in Chapter 1, but what is rangeland? *Rangeland* is land on which the climax plant community is composed principally of grasses, grasslike plants, forbs, and shrubs valuable for grazing and in sufficient quantity to justify grazing use (USDA 1967b).

Rangelands are typically regions of low and erratic precipitation, rough topography, poor drainage, and cold temperatures and are unsuited to cultivation but are a source of forage for free-ranging native and domestic animals as well as a source of wood products and water (Stoddart et al. 1975). Under this definition, forest types that contain a sufficient quantity of forage to graze animals are rangelands. As a result, practically all of the national forests and Bureau of Land Management lands in the western United States have *grazing allotments*, bounded units of land on which grazing is permitted.

3.5.9   GRAZING SYSTEMS

Due to the ecological and economic problems of managing livestock on public lands, a variety of systems has been developed to reduce soil degradation, to reduce competition with wild herbivores, and to provide sustained economic returns. These various systems are termed continuous, deferred, rotation, deferred rotation, and rest rotation (Stoddart et al. 1975). The yearlong use of an area by livestock in consecutive years is referred to as *continuous grazing*. This system reduces livestock handling to a minimum, which in turn reduces labor costs. Moving livestock between different subdivisions of a range is *rotation grazing*. The livestock are moved between pastures on a planned schedule.

Deferred grazing delays grazing for a period of time during seasons when herbaceous plant growth is highest. The idea is to allow time for plants to mature and produce seed before being consumed for food by livestock and wildlife. In the combination system of *rotation-deferment*, grazing is delayed in each pasture, with the deferment rotated on a planned schedule between pastures. *Rest-rotation grazing* is similar to deferred rotation with the exception that the period of deferment for a pasture is an entire year. Other pastures are in turn rested on a rotating basis.

The grazing systems outlined are used as the basis for many combinations of deferment, rotation, and rest specifically tailored to special rangeland conditions. In addition, other systems are available, such as high-intensity, short-duration grazing (Holechek 1983). Each system has certain advantages and disadvantages relative to local conditions that must be evaluated before a grazing system can be selected.

3.5.10   CONFLICTS AND COMPETITION

In the western United States, large numbers of livestock, under the concept of multiple-use management, are permitted to roam free in allotments in the same forests and rangelands occupied by many species of wildlife. Conflicts arise because the diet of some species of livestock overlap the diet of big game, such as mule deer, elk, bighorn sheep, and pronghorn, thereby creating direct competition for food. The range of overlap of plant use between wildlife and domestic species clearly demonstrates the potential for competition (Table 3.6). In general, the food preferences of deer and cattle are sufficiently different that their natural occupancy of a unit is not entirely competitive, while the diets of cattle and elk are almost identical. Discretion must be exercised when comparing overlapping diets so that season as well as location are taken into account.

When several species of grazing animals use the same area, a method has to be devised to compare the use of that ecosystem in equivalency of animals to determine carrying capacity and to provide a basis for charging the allotment holder for the forage eaten by domestic animals. The method that has been adopted is the *animal*

TABLE 3.6
Generalized Overlap of Food Habits of Wildlife and Livestock

Species	Vegetation
	Grass:Forbs:Woody Species
Cattle	************
Bighorn	**********
Horse	*********************
Elk	*************************
Sheep	*****************
Pronghorn	*************
Goats	*****************
Deer	****************

TABLE 3.7
Animal Unit Equivalents for Livestock and Wildlife

Animal (Mature)	Weight, lb (kg)	Forage, lb (kg) Consumed	Animal Unit Equivalents[a]
Bison	800 (818)	36.0 (16.4)	1.80
Bighorn	180 (82)	3.6 (1.6)	0.18
Cattle	1,000 (455)	20.0 (9.1)	1.00
Elk	700 (318)	14.0 (6.4)	0.70
Goat	100 (45)	2.0 (0.9)	0.10
Horse	1,200 (545)	36.0 (16.4)	1.80
Moose	1,200 (545)	24.0 (10.9)	1.20
Mule deer	150 (68)	3.0 (1.4)	0.15
Pronghorn	120 (55)	2.4 (1.1)	0.12
Sheep	150 (68)	3.0 (1.4)	0.15
White-tailed deer	100 (45)	2.0 (0.9)	0.10

Source: Holechek, H. L., *Rangelands* 10:10–14, 1988.

[a] The comparison is cattle with a value of 1.00.

unit: the weight of one mature cow or 1,000 lb (Kothmann 1974). An animal unit is converted to an *animal unit month* (AUM), the amount of forage required for one animal for 1 mo (30 days). To compare the forage consumption between livestock and wildlife, an animal unit equivalent is computed using a 1,000-lb cow consuming 20 lb of forage per day for 30 days or 600 lb (Table 3.7).

The livestock–wildlife problem is not confined to direct competition for food plants but also includes overgrazing of forest and rangeland habitats. *Overgrazing* or overuse refers to the degree of deterioration from a set of original conditions usually comparable to climax grassland. Overgrazing does not always result from too many livestock on a unit but also can result from too many deer and elk. However, control of livestock grazing through animal distribution and manipulation of density is more easily done than controlling wildlife.

Some species of amphibians, birds, small mammals, reptiles, and fishes are also affected by grazing through a loss of cover or special use areas. For example, ground-nesting birds, such as the wild turkey and blue grouse, lose nesting cover and poult- and chick-rearing cover when tall grasses are grazed to stubble in forested areas. Grazing can have a positive effect on wildlife species that depend on brushland for food and cover. The heavy consumption of grasses by livestock leads to the invasion of shrubby species. This happened in the western expansion of the bobwhite quail (Edminister 1954), deer (Holechek 1982), and California quail (Leopold 1977).

The conflicts between wildlife and domestic livestock had their historical beginning in the development of this continent, and some of the problems created in those early years remain today. Overcoming these problems requires the careful integration of social, economical, and ecological factors. Social factors involve how past customs and policies have been carried forward to the present, for example, the policy of issuing grazing permits to one person or family for long periods of time, such as for 99 yr. The holder of the permit has a monopoly on the use of forage on a given piece of land for a specific period of time with the option of permit renewal. The policy is now being questioned by conservation organizations and the general public. However, there is another factor to consider, and this is that ranching is a way of life that has, in many cases, been passed from one generation to another and is still an important economical factor in many parts of the rural West.

Economical factors include low grazing rates charged for the use of forage on federal lands when compared to market value of forage on private lands. In addition, holders of grazing permits on federal lands use them to get loans for which banks consider the acres allotted in the permit to be an extension of the owner's property. Economical factors also include the consumption of red meat as food and how much the public is willing to forgo to recover or maintain existing and future forest and rangeland conditions.

Ecological factors focus on the amount of grazing that can be permitted without damage to soil and watersheds and the loss of forage that otherwise would be available to many species of wildlife, from birds to elk. Indirectly, ecological factors involve not only competition for food and space by livestock and wildlife but also the effects of grazing on nutrient cycling and energy flow that scientists are only now beginning to understand.

3.5.11 NATIONAL RESEARCH COUNCIL

There is no general solution to the livestock-wildlife problems existing today other than complete removal of cattle from public lands. This is not acceptable to ranchers who depend on low grazing fees to maintain a working ranch that provides beef for market. One solution to the problem may be to evaluate each management unit in terms of its capacity to produce goods and services with the objective of protecting the soil base and then open up the use of these goods and services to a competitive process to arrive at a fair market value. Evaluation of management units for their capacity to produce goods and services will require more effort to classify, inventory, and monitor rangelands. The National Academy of Science (NAS 1994) has recommended that the U.S. Departments of Agriculture and Interior cooperate in the following three activities:

- Define and adopt a minimum standard of what constitutes acceptable conditions on rangelands
- Develop consistent criteria and methods of data interpretation to evaluate whether rangeland management is meeting this standard
- Implement a coordinated and statistically valid national inventory to periodically evaluate the health of federal and nonfederal rangelands

The NAS suggests that by implementing these recommendations consistent data and interpretations would be provided to ensure that rangeland ecosystems are protected.

3.6   RESTORE AND MAINTAIN RESOURCES AND DIVERSITY

In the early years of the wildlife profession, managers were largely concerned with population regulation of a small number of game species by controlling the length of hunting seasons and the bag limit. But, with the rapid increase in understanding ecological processes and a growing public and professional concern for maintaining the integrity of ecological systems, wildlife managers rely on cooperative efforts with other natural resource managers to meet the complex planning and management requirements of a large number of species.

Wildlife managers no longer focus exclusively on those animals that hunters find desirable or others find undesirable. They increasingly plan for and manage all wildlife species, many of which are a matter of indifference to a segment of our human population. In view of this approach to management that seeks to benefit all forest wildlife, the following sections focus on best management practices to coordinate with wildlife (3.6.1), timber (3.6.2), range (3.6.3), engineering (3.6.4), watershed (3.6.5), recreation (3.6.6), land use (3.6.7), and aquatic measures (3.6.8). The practices and measures are stated only in a general way; the details must be left to the managers of specific forest types and local conditions.

These details include such considerations as size of openings to create, how much vegetation to leave as cover for a particular species, the amount and species of food plants to seed or plant for a desired wildlife species, and the types and sizes of buffer strips to leave along streams. These and other factors have to be tailored to a particular vegetation type and given species. Any management practice or coordination measure used must be closely aligned to the silvicultural requirements of tree species in the forest type and the life history requirements of the wildlife species associated with the type.

Most of the direct habitat management practices listed are the result of research or many years of practical experience. The list is not all inclusive, but it does contain a number of generic measures suitable for a variety of forest types. Detailed descriptions of each practice can be found in many state and federal publications dating from the 1930s to the present. The practices listed are for the most part deliberately undertaken to benefit wildlife, so all the costs are usually paid from wildlife funds. There are many practices that can be used to improve wildlife habitat in forested ecosystems. However, before transporting technology from one forest region to another, it is important to document that the practice actually benefits the targeted species or group of species.

Many wildlife habitat practices and coordination measures require the seeding, planting, or protection of areas containing certain trees, shrubs, or forbs as wildlife *food* or *cover* plants. Plant species possess nutritional elements in differing quantities, and their value for one wildlife species may be different from that for another. On the other hand, a single plant species may possess food or cover value for a number of wildlife species. Clearly, such plants are those that managers should preferentially employ in habitat improvement projects.

Single live *trees* provide the physical structure needed for a nest by a squirrel or a hawk, for overnight roosting by wild turkeys, or for perching and resting by doves and pigeons. When single trees accumulate to make a small group and can be identified as a stand, then additional protection is provided to many forms of wildlife. In their early developmental stage, sprouts, limbs, or twigs of trees are food for browsing animals such as deer and elk. The importance of live trees in providing wildlife food and cover cannot be overemphasized. Approximately 28 genera make up the groups of tree species across North America that are commonly used by some wildlife species (Table 3.8).

Forests are more than live, straight trees used for sawlog production. Dead trees, snags, hollow trees, decaying logs, and stumps are used as hunting perches for raptors; as den trees for raccoons, squirrels, and bears; as feeding sites for woodpeckers;

TABLE 3.8
North American Trees Used by a Number of Wildlife Species for Food and Cover

Common Name	Scientific Name (Genus)	Common Name	Scientific Name (Genus)
Alder	*Alnus*	Hickory	*Carya*
Ash	*Fraxinus*	Incense-cedar	*Libocedrus*
Aspen	*Populus*	Juniper	*Juniperus*
Baldcypress	*Taxodium*	Larch	*Larix*
Beech	*Fagus*	Magnolia	*Magnolia*
Birch	*Betula*	Maple	*Acer*
Blackgum	*Nyssa*	Mulberry	*Morus*
Cedar	*Thuja*	Oak	*Quercus*
Cherry	*Prunus*	Pine	*Pinus*
Chinaberry	*Melia*	Sassafras	*Sassafras*
Dogwood	*Cornus*	Sourwood	*Oxydendrum*
Douglas-fir	*Pseudotsuga*	Spruce	*Picea*
Fir	*Abies*	Willow	*Salix*
Hemlock	*Tsuga*	Tuliptree	*Liriodendron*

and as nests for cavity-nesting birds. On occasion, rabbits use rotted tree root canals for tunnels and dens, as do snakes, lizards, and small rodents. The cavity-nesting birds are a particularly important class of forest birds because the majority of them are insectivorous and help control endemic forest insects that damage valuable timber trees. The process from a seed to live tree and later to a snag is an ecological trail showing the importance of understanding ecological systems in managing forests for wildlife. An example using ponderosa pine illustrates the point:

> A seed from the cone of a ponderosa pine falls to a moist spot on the ground, sprouts roots, and starts to grow. In a few years, the seed has become a seedling 8–10 in high with a diameter of 1/4 in. From this early beginning, the seedling progresses through development stages until it reaches a diameter of 4 in, when by definition it becomes a tree. The tree continues to grow through a sapling and intermediate stage to a mature tree with a diameter of 20 in. At this seasoned age of 200 yr, the tree has developed a yellow color and platy bark with deep fissures. During the growth period from a sapling to a mature tree, wind, lightning, mistletoe, and bark beetles are at work affecting the health of the tree and reducing its resistance to the slow process of decay and death. As a live tree, this ponderosa pine has probably provided homes for insects, which are food for birds such as the pygmy nuthatch and brown creeper. It also may have provided cover for a squirrel nest, and there could have been bats or a nest for a brown creeper in the small space between the loose bark plates. Depending on its growth form (density of the crown and size of branches), the tree could have been used by turkeys for roosting and provided nest sites for hawks and owls, and deer would have eaten mistletoe from its branches near the ground.

A live tree provides food, homes, and special use sites for many species of wildlife, but use continues when the tree is dead. Just as a live tree goes through development stages, a dying tree goes through decaying stages from having a solid trunk, with brown needles for leaves, to the final stage during which most of the bark and branches are gone and the trunk has become soft and easy to penetrate.

Across North America, snags are used by approximately 85 species of birds, over 45 species of mammals, as well as an undetermined number of species of reptiles, amphibians, and invertebrates (Table 3.9). In the Southwest, soft snags larger than 15 in are used by 53 species of primary and secondary cavity-nesting birds. Primary cavity nesters (PCNs) (woodpeckers and nuthatches) excavate holes in the soft decaying wood. They make new holes each year, leaving their old ones to be used again by other wildlife. Holes and fresh chips at the base of trees are signs that the tree is used by cavity nesters. Secondary cavity nesters (SCNs) (owls, kestrels, bluebirds, wood ducks, flying squirrels) cannot excavate their own holes and so make their homes in those abandoned by PCNs and natural cavities created by broken tops, lost branches, and lightning strikes.

Cavity-nesting birds are mostly insectivorous and help control insect populations that may kill trees before they reach maturity. Snags larger than 20-in diameter are used more frequently by cavity nesters. There are several reasons why large snags are the most beneficial: They provide more space for birds needing larger cavities, nest insulation is better, they stand longer, and more substrate is available

TABLE 3.9

Some Wildlife Species that Use Dead Trees and Snags for Food and Cover

Common Name	Life Form	Common Name	Life Form
Acorn woodpecker	bi	Mountain bluebird	bi
Ash-throated flycatcher	bi	Northern flying squirrel	ma
Black-backed woodpecker	bi	Northern raccoon	ma
Black-capped chickadee	bi	Porcupine	ma
Brown-headed nuthatch	bi	Prothonotary warbler	bi
Carolina chickadee	bi	Pygmy nuthatch	bi
Douglas's squirrel	ma	Three-toed woodpecker	bi
Downy woodpecker	bi	Vaux's swift	bi
Eastern fox squirrel	ma	Western bluebird	bi
Eastern screech-owl	bi	Western gray squirrel	ma
Hairy woodpecker	bi	Wood duck	bi
House wren	bi	Yellow-bellied sapsucker	bi
Ladder-backer woodpecker	bi		

Note: Selected from the Life-Needs Relation in the FAAWN data model.
bi, bird; ma, mammal.

for insects that are a source of food. Large snags (28–30 in) are essential for species such as the pygmy nuthatch, which roosts communally to conserve heat energy. In one instance, over 160 birds simultaneously used one tree cavity. Brown creepers and Williamson's sapsuckers also require large snags. The following are some uses of snags by wildlife in the Southwest:

- Open nests: osprey, bald eagle, and heron
- Cavity nests: goldeneye, woodpeckers, owls, and chickadees
- Hunting perches: hawks and flycatchers
- Food storage: squirrels and jays
- Food and shelter: chickadees and bats
- Resting and roosting: swallows, hawks, nuthatches

The number of dead trees, snags, logs, and stumps to leave as a component of wildlife habitat is variable depending on the species and forest region. For example, in Arizona and New Mexico the U.S. Forest Service policy is to leave three per acre within 500 ft of forest openings and water, with two per acre over the remaining forest (USDA 1976). In the Southwest, good snags are considered to be over 6 years old, more than 18-in dbh (diameter at breast height), and with more than 40% bark cover. In the Pacific Northwest, the suggested snag density is six to seven per acre for ponderosa pine (Thomas et al. 1979). The wildlife species using dead trees and snags in the seven forest regions include more than cavity-nesting birds, as Table 3.9 indicates.

TABLE 3.10
Shrubs Used by a Number of Wildlife Species for Food and Cover

Common Name	Scientific Name (Genus)	Common Name	Scientific Name (Genus)
Apache-plume	*Fallugia*	Juniper	*Juniperus*
Azalea	*Rhododendron*	Kalmia	*Kalmia*
Barberry	*Berberis*	Manzanita	*Arctostaphylos*
Bearmat	*Chamaebatia*	Menziesia	*Menziesia*
Blueberry	*Vaccinium*	Mountainmahogany	*Cercocarpus*
Buckthorn	*Rhamnus*	Oak	*Quercus*
Bush-honeysuckle	*Diervilla*	Palmetto	*Sabal*
Buttonbrush	*Cephalanthus*	Paperflower	*Psilostrophe*
Calliandra	*Calliandra*	Porliera	*Porliera*
Ceanothus	*Ceanothus*	Rabbitbrush	*Chrysothamnus*
Cherry	*Prunus*	Raspberry	*Rubus*
Chokecherry	*Aronia*	Rose	*Rosa*
Cinquefoil	*Potentilla*	Sagebrush	*Artemesia*
Cliffrose	*Cowania*	Saw palmetto	*Serenoa*
Condalia	*Condalia*	Serviceberry	*Amelanchier*
Cyrilla	*Cyrilla*	Silktassel	*Garrya*
Dalea	*Dalea*	Snowberry	*Symphoricarpos*
Elder	*Sambucus*	Sumac	*Rhus*
Euonymus	*Euonymus*	Viburnun	*Viburnum*
Gooseberry	*Ribes*	Waxmyrtle	*Myrica*
Hackberry	*Celtis*	Willow	*Salix*
Hawthorn	*Crataegus*	Winterfat	*Eurotia*
Hazelnut	*Corylus*	Witchhazel	*Hamamelis*
Holly	*Ilex*	Wolfberry	*Lycium*
Huckleberry	*Gaylussacia*		

Planting and production of *shrubs* was emphasized in the past, particularly in the West, due to their importance in maintaining big game populations. More recent ecological thinking has encouraged their cultivation for a variety of other life forms as well. Many shrubs not only are browse food plants for deer and elk but also provide seeds and fruits for birds and small mammals and, in many cases, provide cover for the same groups of species (Table 3.10). Approximately 1,000 species of shrubs, semishrubs, and woody vines grow in the U.S. national forests and adjacent lands (Dayton 1931). Martin, Zim, and Nelson (1951), in their classic book, account for over 100 wildlife species using shrubs for food.

Herbaceous plants include the many species of grasses, such as wheatgrasses, bluegrasses, and bromes and the low, wide-leaved flowering plants, such as buttercups, cinquefoils, and clovers, often referred to as forbs. Grasses, grasslike plants, and herbaceous plants are important sources of wildlife food and provide cover for a range of animals from small mammals such as chipmunks, mice, and ground squirrels, to large grazing animals, including elk, deer, and pronghorn. Given limits of

TABLE 3.11

Forbs Used by a Number of Wildlife Species for Food and Cover

Common Name	Scientific Name (Genus)	Common Name	Scientific Name (Genus)
Agoseris	*Agoseris*	Medic	*Medicago*
Alumroot	*Heuchera*	Oxalis	*Oxalis*
Arnica	*Arnica*	Parthenium	*Parthenium*
Balsamroot	*Balsamorhiza*	Partridgeberry	*Mitchella*
Brodiaea	*Dichelostemma*	Peavine	*Lathyrus*
Bundleflower	*Desmanthus*	Penstemon	*Penstemon*
Buttercup	*Ranunculus*	Plantain	*Plantago*
Cinquefoil	*Potentilla*	Pussytoes	*Antennaria*
Clover	*Trifolium*	Ragweed	*Ambrosia*
Croton	*Croton*	Royal fern	*Osmunda*
Dandelion	*Taraxacum*	Senna	*Cassia*
Deervetch	*Lotus*	Snapweed	*Impatiens*
Dock	*Rumex*	Strawberry	*Fragaria*
Eriophyllum	*Eriophyllum*	Sweetclover	*Melilotus*
Eupatorium	*Eupatorium*	Tansymustard	*Descurainia*
Filaree	*Erodium*	Thistle	*Cirsium*
Fleabane	*Erigeron*	Tickclover	*Desmodium*
Galax	*Galax*	Trailing arbutus	*Epigaea*
Goldenrod	*Solidago*	Turkeymullein	*Eremocarpus*
Groundcherry	*Physalis*	Vetch	*Vicia*
Groundsmoke	*Gayophytum*	Violet	*Viola*
Hawkweed	*Hieracium*	Waterlily	*Nymphaea*
Hollyfern	*Polystichum*	Wild-buckwheat	*Eriogonum*
Knotweed	*Polygonum*	Willow-herb	*Epilobium*
Lespedeza	*Lespedeza*	Wintergreen	*Gaultheria*
Lettuce	*Lactuca*	Wyethia	*Wyethia*
Lupine	*Lupinus*	Yarrow	*Achillea*
Marshpurslane	*Ludwigia*		

space, it is not possible here to list and describe all the grass or forb species of value to all wildlife species, but an appreciation of their importance can be obtained by reading current issues of the *Wildlife Society Bulletin* or *Journal of Wildlife Management*. Articles in these journals contain information on herbaceous plants used by various wildlife species for either food or cover. Some of the most common genera are listed in Tables 3.11 and 3.12.

3.6.1 WILDLIFE COORDINATION AND RESTORATION MEASURES

The opportunity to improve conditions for wildlife through indirect techniques or restoration measures with other agencies can often be accomplished at greatly reduced costs and with a minimum expenditure of wildlife funds. Indirect techniques often

TABLE 3.12

Grass and Grasslike Plants Used by a Number of Wildlife Species for Food and Cover

Common Name	Scientific Name (Genus)
Grass	
Bluegrass	*Poa*
Bristlegrass	*Setaria*
Brome	*Bromus*
Cane	*Arundinaria*
Cupscale	*Sacciolepis*
Cutgrass	*Leersia*
Fescue	*Festuca*
Junegrass	*Koleria*
Muhly	*Muhlenbergia*
Needlegrass	*Stipa*
Panicum	*Panicum*
Paspalum	*Paspalum*
Saltgrass	*Distichlis*
Squirreltail	*Sitanion*
Wheatgrass	*Agropyron*
Wild rye	*Elymus*
Grasslike	
Bulrush	*Scirpus*
Rush	*Juncus*
Sedge	*Carex*
Spikerush	*Eleocharis*

depend on coordination of activities with other land management agencies, including those dealing with timber, range, engineering (roads, etc.), watershed, and recreation as well as management of other special land use projects.

Restoration measures differ from direct habitat management practices only in the way that they are financed and brought to completion. These measures generally use funds from several resource agencies or functions to accomplish the job. Restoration measures like those in the following list are the bread and butter of wildlife management, and their contribution should not be underestimated even though results in terms of increased wildlife populations may not be immediately evident.

1. Maintain (plant, cultivate, and fertilize) permanent openings (food plots) with wildlife food plants. This practice benefits many species but generally is used to attract deer and turkey to specific areas.
2. Cut, burn, or disk concentrations of shrub species to stimulate growth of browse. These techniques are used primarily to produce food for deer and elk.

3. Create snags by directly killing trees. This benefits many species of cavity-nesting birds and provides hunting perches for raptors.
4. Fence riparian areas to exclude cattle, which benefits many species, but cost is high.
5. Restrict access to critical habitat of threatened and endangered species. This is a proven technique, but cost may exclude its use in many areas. This provides a good public involvement project for volunteers.
6. Plant food and cover species along streams to stabilize bank erosion and provide shade. Generally, this is a good conservation technique that will benefit aquatic and terrestrial species.
7. Create water sources by building small ponds, guzzlers, and trick tanks. This will benefit many species. Cost can be high, and maintenance is required.
8. Improve existing water sources such as seeps and springs to create permanent water sources. The area may need to be fenced to exclude livestock.
9. Build dams to create shallow lakes for waterfowl habitat. This is a good technique but can be expensive. The best situation is for a multiresource structure as in flood control.
10. Build small check dams in eroded wet meadows. Generally, this provides quick results and conditions to establish riparian vegetation.
11. Remove invading trees from wet meadows. This is expensive but effective in maintaining high-forage-producing areas for deer and elk and "bugging areas" for turkey poults.

3.6.2 Timber Coordination and Restoration Measures

Timber management activities affect a large number of wildlife species either positively or negatively depending on the requirements of the individual species. For this reason, the removal of tree overstory can be used as a management tool to meet specific wildlife objectives. The coordination measures listed here are intuitive but are based on sound ecological principles:

1. Locate timber roads and skid trails to eliminate or minimize siltation of streams.
2. Reserve areas for food and cover, in connection with tree planting and regeneration, natural openings, access ways, and brush. This benefits deer, elk, and turkey.
3. Plant or reserve groups of trees for cover (thermal, escape, hiding, etc.) for targeted species.
4. Reserve and release fruit and nut trees and shrubs to increase food production. This benefits many species.
5. Retain den and roost trees. Generally, this is species specific for squirrels, raccoons, turkeys, band-tailed pigeons, and some raptors.
6. Create openings by harvesting in dense timber stands. This is one of the most common techniques to benefit deer, elk, and turkey as well as many species of small mammals and birds.

7. Plan timber harvest roads to provide access for hunters and people who fish.
8. Retain tree buffer areas or riparian vegetation along streams and lakeshores to maintain cold water temperatures.
9. Include provisions in timber sale contracts to prevent entry of large amounts of small-size debris into perennial streams. Debris of large size may be desirable in some streams.
10. Favor aspen reproduction in timber sales where it is a seral stage. The management objective must be clearly defined, especially where beaver or deer could cause damage to a watershed.
11. Locate logging roads away from live streams to prevent destruction of natural stream channels. Often, this requires persistence to keep roads away from water.
12. Prohibit log landings in live streams in timber sale contracts.
13. Bridge water courses on log-haul roads.
14. Limit the number of stream crossings in logging operations. This may increase cost of road building to an unacceptable level.
15. Eliminate tree felling into stream channels or dragging trees along or across channels. Enforcement and supervision is a must where sale affects an important fishery.
16. Prevent stream pollution from sawmill waste.
17. Adjust slash disposal plans to include requirements of wildlife species; slash piles benefit small mammals and birds.
18. Prevent log skidding across meadows, along stream banks or through food, escape, nesting, or roosting cover. Areas must be identified before the sale.
19. Reserve cover adjacent to seeps and stringer meadows in cut areas. Areas must be identified before timber sale.
20. Promote sales to regenerate aspen. This is an effective technique to increase forage production for deer and elk where the type occurs.
21. Withhold from tree planting special areas needed for food and cover plantings. Designate species to benefit in management plans.
22. In releasing coniferous timber species, leave hardwoods in strips or patches to provide a mast crop. This benefits squirrel, deer, and turkey.
23. Suspend logging during spawning seasons to avoid interference with upstream movement of redds and fry.
24. Scatter the location of small timber sales to break up large areas of a single age class. This provides diversity of food and cover areas for many species.
25. In large clearcuts, leave small scattered plots uncut for cover. This increases use of large cut areas by deer and elk.
26. Seed skidways, roadsides, and landings on sale areas with wildlife food mixtures of forbs, grasses, and shrubs. This provides food for many species.
27. Burn slash away from water so ash will not enter streams or lakes.
28. Provide for juxtaposition of stands in different vegetation types.
29. Provide for interspersion of habitat units throughout the vegetation type.

30. Thin dense stands of trees to stimulate understory growth. This is a good technique for deer and elk.
31. Close permanent and seasonal logging roads when necessary to protect a species during all or part of a year.
32. Mark and save a given number of snags per acre for cavity-nesting birds.

3.6.3   RANGE COORDINATION AND RESTORATION MEASURES

The coordination measures undertaken with range management agencies are largely directed toward including wildlife requirements for grazing and browsing species such as deer, elk, and bighorn sheep in allotment analysis plans and reducing the livestock-wildlife competition for food and space. The following coordination measures are those that range conservationists and wildlife biologists have generally agreed on due to their wide application. They are practical and beneficial and can be accomplished with little difficulty.

1. Allocate a percentage of forage to wildlife species in range allotment plans to keep total livestock-wildlife numbers within carrying capacity.
2. Manage livestock to protect key wildlife areas. This is especially important in overlapping winter use areas.
3. In placing salt on the range, distribute it to allow use by big game animals. While this technique was used extensively in the past, there is some question about its value for wildlife.
4. Reseed depleted grass ranges with palatable species to reduce livestock use of shrubs. Planting is done according to local guidelines.
5. Allow grazing of moderate intensity for shorter periods to reduce livestock use of shrubs.
6. Install escape ramps for small game in livestock watering tanks to keep water from being polluted by dead animals.
7. Include legumes, browse, and other important wildlife foods in range-reseeding mixtures. Plant species adapted to local conditions and palatable to target species such as deer, elk, and turkey.
8. Leave browse for wildlife in reseeding projects involving brush removal. Make certain browse species are not decadent.
9. Defer spring livestock grazing until young wildlife are no longer dependent on green food. This is effective in important game bird habitat and also benefits deer and elk.
10. Control, fence, or eliminate livestock driveways across winter game ranges. Fencing is expensive.
11. When initiating brush control on winter ranges, consider wildlife needs in the early planning stages of a project in relation to the surrounding area.
12. Consider the possible competition effects on wildlife when changing the kind of livestock in an allotment. Competition between livestock and wildlife must be considered early in the planning process.

13. Reserve forage at heavily used hunting and fishing camps for recreational pack and saddle stock.
14. Do not build range improvements that concentrate or encourage use of key wildlife habitat by domestic livestock. Plan for a good distribution of improvements.
15. Locate stock-watering improvements to benefit wildlife. Water improvements have proved an important factor in increasing deer, elk, and turkey where food and cover, but not water, were available.
16. Fence springs at water developments to provide cover and nesting sites for upland birds. This is an effective technique in heavily stocked livestock allotments.
17. Consider the effects on other wildlife when initiating predator control, particularly if poison is used.
18. Where there is direct competition for forage between livestock and big game, select a reasonable balance. This must be provided for in the planning stage.

3.6.4   ENGINEERING COORDINATION AND RESTORATION MEASURES

Engineering concerns deal primarily with cultural and structural features, including roads. The placement of buildings, fences, recreational areas, electrical lines, and water structures can have highly detrimental effects on many wildlife species. As a consequence, coordination measures undertaken with engineering agencies are largely targeted at mitigating negative impacts on habitat and populations. The following measures are among those wildlife biologists have found to be most acceptable and successful:

1. Limit borrow pits from gravel bars to above water level. Borrow pits below the high-water stage should be leveled and provided with drainage to avoid trapping fish when high water recedes.
2. Install culverts in streams at the grade of the stream and below stream level. Drop-offs at outlets are impossible for fish to negotiate. Drop-offs may undercut culverts and require expensive maintenance procedures.
3. Eliminate or minimize use of heavy equipment in streambeds to prevent silting of streams.
4. Consider the access needs for hunters and those who fish when planning roads and trails.
5. In locating roads, avoid stream banks, meadows, water courses, and other areas used by wildlife. Locating roads away from water or wet areas is also a good soil conservation technique.
6. Plant wildlife food on cut-and-fill slopes. This is a good conservation practice to reduce erosion. Plant species palatable to targeted wildlife.
7. For roads along a stream, leave fringes of timber and brush to provide shade and bank protection.
8. Keep construction debris out of streams to reduce siltation of spawning areas.

9. Provide for spreading of drainage water from roads to prevent silting of streams. This reduces siltation and pollution.
10. Consider designing road fill areas to serve as dams that can provide additional water. This must be planned carefully so that the dirt dam is stable.

3.6.5 WATERSHED COORDINATION AND RESTORATION MEASURES

Considerable watershed restoration work is done under state and federal programs to provide jobs for unemployed people. Such projects are often dependent on the political climate, so when a favorable climate exists, a backlog of projects can often be completed in a short time. Some of the measures listed grew out of the experience of the Civilian Conservation Corp of the 1930s but remain just as effective today. In the 1980s, these coordination measures were often programmed for specific projects in small watersheds with small work crews.

1. Use browse and legumes in erosion control plantings to benefit wildlife species.
2. Include wildlife food plants in bank stabilization projects.
3. Integrate watershed plans and big game hunting to provide for game management in municipal watersheds to increase recreational opportunities for a local area.
4. Include the protection and improvement of wildlife habitat in water impoundment programs. If planned properly, this can benefit many locally important species.
5. Safeguard or improve aquatic habitats when using structures to alter stream courses or stream banks or for channel and gully control. This is a commonsense practice.
6. In gully and other erosion control plantings, use fruit-bearing and thicket-forming species to provide food and cover for wildlife.

3.6.6 RECREATION COORDINATION AND RESTORATION MEASURES

Recreation coordination measures are aimed at controlling the behavior of people, not only hikers, picnickers, and the like, but hunters and fishermen as well. Considerable effort is required to mitigate the effects of recreational activities on both habitat and populations, particularly on sites having fragile soils or containing sensitive plants and animals. Most recreation control measures require long-range planning involving public review and acceptance.

1. In sensitive soil areas, restrict cross-country use of four-wheel drive vehicles or all-terrain vehicles to designated roads. This is a controversial technique that always requires public review.
2. Locate hunting and fishing camps and enforce camping regulations so that stream pollution or damage to habitat is held to a minimum.
3. Corral horses used at hunting camps located within deer and elk winter ranges. Enforcement may be difficult in remote areas.

4. Restrict camping and grazing of pack and saddle animals at heavily used hunting and fishing sites. The terrain immediately adjacent to these areas may need to be closed to grazing.
5. Restrict speed boating in high-quality fishing waters.
6. Consider camping needs of hunters and fishermen when selecting areas for campground development.
7. Through hunting and fishing clubs and other organizations, develop programs of cooperative camp cleanup, sanitation, and fire prevention. This is an effective way to educate sports enthusiasts on just how much trash is left by their activities.
8. Plan for the needs of noncommercial hunters and fishermen for use of campsite areas before issuing special permits to commercial guides.

3.6.7 Land Use Coordination and Restoration Measures

The coordination measures for land use grow out of public and private projects not normally part of the day-to-day management of wildlife. Such projects often result from state or federal undertakings that create or maintain large bodies of water or extensive land areas or that provide benefits to large numbers of people at a distance from the management area. As with the other coordination measures, the following list is not inclusive:

1. Install fish screens at head gates of irrigation ditches.
2. Maintain minimum pool levels in artificial reservoirs as a mitigating procedure. This must be provided for early in the planning stage.
3. Keep minimum and maximum rates of discharge from impoundments below levels hazardous to fish and wildlife. Determining such rates must be done early in the planning process.
4. Provide vegetative cover on ditch banks to reduce soil erosion.
5. Consider needs of wildlife when acting on applications for agricultural use of land.
6. Provide for protection of wildlife and habitat, including the control of organic and inorganic pollution, when issuing occupancy permits. Enforcement of existing regulations is generally the major problem.
7. Plant forb, grass, and browse species along power lines and similar clearings to produce wildlife food and cover and at the same time reduce the cost of periodic clearing. If properly coordinated with utility companies, this technique can be effective in providing food and cover for many wildlife species.

3.6.8 Aquatic Coordination and Restoration Measures

Restoration, sustaining, and other measures are just as important to maintaining fish species as they are for terrestrial species. As with birds, mammals, and reptiles, habitat improvement is best done through coordination measures where funds can be combined from several sources to accomplish multiple management objectives. Some direct measures that have been successful in improving aquatic resources are as follows:

1. Fertilize ponds to increase phytoplankton as a food source with a 5–10–5 fertilizer but use caution to prevent amounts that can cause pollution.
2. Plant trees or tall shrubs along stream banks to maintain a cold water temperature.
3. Build small dams in streams to create pools. This is an excellent stream improvement technique. Dam maintenance must be included in future work plans.
4. Create gravel beds in lakes for spawning areas. This is a cost-effective technique, but it requires review and planning to verify what is needed.
5. Create cover in lakes by sinking brush piles or old car bodies. This is an effective technique for providing cover at a low cost.
6. Build small fishing lakes within the forest environment. This requires planning for location and degree of use.
7. Remove undesirable vegetation from lakes and ponds. Heavy concentrations of aquatic vegetation can destroy a good local fishery.

Habitat improvement to maintain a good trout stream is expensive. Improvements in the stream must be preceded by improvements in watershed conditions to achieve long-term results.

KNOWLEDGE ENHANCEMENT READING

Barbour, M. G., and D. W. Billings. 2000. *North American terrestrial vegetation.* Cambridge University Press, Cambridge, UK.

Fulbright, T. E., and D. G. Hewitt (Eds.). 2008. *Wildlife science, linking ecological theory and management applications.* CRC Press, Taylor and Francis Group, Boca Raton, FL.

Gibbs, J. P., M. J. Hunter, and E. L. Sterling. 2008. *Problem-solving in conservation biology and wildlife management.* Blackwell, Malden, MA.

Gilbert, F. G., and D. G. Dodds. 2001. *The philosophy and practice of wildlife management.* Krieger, Malabar, FL.

Morrison, M. L., B. G. Marcot, and R. W. Mannan. 1998. *Wildlife-habitat relationships: Concepts and applications.* University of Wisconsin Press, Madison.

Nyland, R. D. 2002. *Silviculture concepts and applications.* McGraw-Hill, New York.

Smith, D. M., B. C. Larson, M. J. Kelty, et al. 1997. *The practice of silviculture: Applied forest ecology.* John Wiley and Sons, New York.

Tappeiner, J. C., II, D. A. Maguire, and T. B. Harrington. 2007. *Silviculture and ecology of western U.S. Forests.* Oregon State University Press, Corvallis.

Walker, L. C. 1999. *The North American forests: Geography, ecology and silviculture.* CRC Press, Taylor and Francis Group, Boca Raton, FL.

Wenger, K. F. (Ed.). 1984. *Forestry handbook*, 2nd ed. The Society of American Foresters. John Wiley and Sons, New York.

REFERENCES

Anderson, R. C. 1972. The ecological relationships of meningeal worm and captive cervids in North America. *J. Wildl. Dis.* 8:304–310.

Anderson, R. C., and A. K. Priestwood. 1981. Lungworms. In W. R. Davidson (Ed.), *Diseases of white-tailed deer*, pp. 266–317. Miscellaneous Publication 7. Tall Timbers Research Station, Tallahassee, FL.

Baker, F. S. 1949. A revised tolerance table. *J. Forest.* 47:179–181.

Blockstein, D. E. 1990. Toward a federal plan for biological diversity. *J. Forest.* 88:15–19.

Brooks, F. C. 1971. Senate hears clearcutting concerns. *J. Forest.* 69:299–302.

Cahalane, V. H. 1939. Integration of wildlife management with forestry in the Central states. *J. Forest.* 7:162–166.

Chapman, H. H. 1936. Forestry and wildlife. *J. Forest.* 34:104–106.

Conant, R., and J. T. Collins. 1998. *A field guide to reptiles and amphibians.* Houghton Mifflin, Boston.

Copp, J. D. 1975. Why hunters like to hunt. *Psychol. Today* 12:65–67.

Davenport, L. A. 1940. Timber vs. wildlife. *J. Forest.* 38:661–666.

Davidson, W. R., F. A. Hayes, V. F. Nettles, and F. E. Kellog (Eds.). 1981. *Diseases and parasites of white-tailed deer.* Tall Timbers Research Station, University of Georgia, Athens.

Dayton, W. A. 1931. *Important western browse plants.* Miscellaneous Publication 101. U.S. Department of Agriculture, Washington, DC.

Edminister, F. C. 1954. *American game birds of field and forest.* Charles Scribner's Sons, New York.

Forbs, R. D. (Ed.). 1961. *Forestry handbook*, Forest wildlife management, Working Group, Section 9, pp. 38–39. Ronald Press, New York.

Fox, R., and B. Rhea. 1989. Spotted owls in the Rockies. *J. Forest.* 87:41–45.

Gabrielson, I. N. 1936. The correlation of forestry and wildlife management. *J. Forest.* 34:98–103.

Giles, R. H., Jr., and L. A. Nielson. 1990. A new focus for wildlife resource managers. *J. Forest.* 88:21–26.

Gross, T. E., and D. P. Dykstra. 1989. Harvest scheduling allowing well-defined violations to age class non-adjacency constraints. In A. Techle, W. W. Covington, and R. H. Hamre (Eds.), *Multiresource management of ponderosa pine forests*, pp. 165–175. General Technical Report RM-185. U.S. Forest Service, Fort Collins, CO.

Harris, L. D. 1984. *The fragmented forest: Island biogeography theory and the preservation of biological diversity.* University of Chicago Press, Chicago.

Helms, J. A. (Ed.). 1998. *The dictionary of forestry.* Society of American Foresters, Bethesda, MD.

Holechek, H. L. 1982. Managing rangelands for mule deer. *Rangelands* 4:25–28.

Holechek, H. L. 1983. Considerations concerning grazing systems. *Rangelands* 5:208–211.

Holechek, H. L. 1988. An approach for setting the stocking rate. *Rangelands* 10:10–14.

Holt, S. J., and L. M. Talbott. 1978. *New principles for the conservation of wild living resources.* Wildlife Monograph 59. Wildlife Society, Washington, DC.

Hooven, E. F. 1973. A wildlife brief for the clearcut logging of Douglas-fir. *J. Forest.* 71(4):210–214.

Jacobson, H. A, D. C. Guynn, and E. J. Hackett. 1979. Impact of the botfly on squirrel hunting in Mississippi. *Wildl. Soc. Bull.* 7:46–48.

Kellert, S. R. 1980. Americans' attitudes and knowledge of animals. *Trans. N. Am. Wildl. Nat. Resources Conf.* 43:412–423.

Kothmann, M. (Ed.). 1974. *A glossary of terms used in range management.* Society of Range Management, Denver, CO.

Kramp, B., W. Brady, and D. R. Patton. 1983. *The effects of fire on wildlife habitat and species*. Wildlife Unit Technical Report. U.S. Forest Service SW Region, Albuquerque, NM.

Lee, R. G. 1990. The redesigned forest. *J. Forest*. 88:32–34.

Leopold, A. 1933. *Game management*. Charles Scribner's Sons, New York.

Leopold, A. S. 1977. *The California quail*. University of California Press, Berkeley.

Lyon, L. J., J. Crawford, H. Czuhai, et al. 1978. *Effects of fire on fauna: A state-of-knowledge review*. Washington Office. General Technical Report 6. U.S. Forest Service, Washington, DC.

Martin, A. C., H. S. Zim, and A. L. Nelson. 1951. *American wildlife and plants: A guide to wildlife food habits*. Dover, New York.

Mealey, S. P., J. F. Lipscomb, and K. N. Johnson. 1982. Solving the habitat dispersion problem in forest planning. *Trans. N. Am. Wildl. Nat. Resources Conf*. 47:142–153.

National Academy of Science. 1994. *Rangeland health: New methods to classify, inventory and monitor rangelands*. National Research Council, Washington, DC.

Peterson, R. T. 1980. *A field guide to eastern birds*. Houghton Mifflin, Boston.

Peterson, R. T. 1990. *A field guide to western birds*. Houghton Mifflin, Boston.

Petrides, G. A. 1988. *A field guide to eastern trees*. Houghton Mifflin, Boston.

Petrides, G. A. 1992. *A field guide to western trees*. Houghton Mifflin, Boston.

Reid, F. A. 2006. *Mammals of North America*. Houghton Mifflin, Boston.

Resler, R. A. 1972. Clearcutting: Beneficial aspects for wildlife resources. *J. Soil Water Cons*. 27:250–254.

Schmidt, R. L., C. P. Hibler, T. R. Spraker, et al. 1979. An evaluation of drug treatment for lungworm in bighorn sheep. *J. Wildl. Manage*. 43:461–467.

Smith, R. H., D. J. Neff, and N. G. Woolsey. 1986. Pronghorn response to coyote control: A benefit/cost analysis. *Wildl. Soc. Bull*. 14:226–231.

Stebbins, R. C. 2003. *A field guide to western reptiles and amphibians*. Houghton Mifflin, Boston.

Stoddart, L. A., A. D. Smith, and T. W. Box. 1975. *Range management*. McGraw-Hill, New York.

Thomas, J. W., R. G. Anderson, C. Maser, et al. 1979. Snags. In J. W. Thomas (Ed.), *Wildlife habitats in managed forests: The Blue Mountains of Oregon and Washington*. Agriculture Handbook 553. U.S. Forest Service, Washington, DC.

U.S. Department of Agriculture (USDA). 1967a. *Major forest types* [map]. National Atlas. U.S. Forest Service, Washington, DC.

U.S. Department of Agriculture (USDA). 1967b. *National handbook for range and related grazing lands*. Soil Conservation Service, Washington, DC.

U.S. Department of Agriculture (USDA). 1976. *Fire management*. Title 5100. U.S. Forest Service Region 3 Supplement 19. Albuquerque, NM.

U.S. Department of Agriculture (USDA). 1989. *An analysis of the land base situation in the United States: 1989–2040*. General Technical Report RM-181. U.S. Forest Service, Fort Collins, CO.

U.S. Department of Agriculture (USDA). 2000. *Forest cover types* [map]. The National Atlas of the United States of America. Compiled by U.S. Forest Service, Washington, DC, and U.S. Geological Survey, Denver, CO.

Walters, C. 1986. *Adaptive management of renewable resources*. Macmillan, New York.

Wood, G. W. 1990. The art and science of wildlife management. *J. Forest*. 88(3):8–12.

Wright, H. A., and A. W. Bailey. 1982. *Fire ecology, United States and southern Canada*. John Wiley and Sons, New York.

Yeatter, R. E., and D. H. Thompson. 1952. Tularemia, weather, and rabbit populations. *Ill. Nat. Hist. Survey. Bull*. 25:351–382.

4 Management Strategies

4.1 INTRODUCTION

What wildlife habitat can a healthy ecosystem produce over time that can be used by the current generation and still meet the requirements of forest sustainability (Figure 1.1)? This question is more philosophical than real, with the expectation that reality will occur as a result of new information to set direction and policy. Long-term sustainability of forest habitats can be accomplished by adopting the position of ecosystem management and wisely using adaptive resource management to accomplish goals and objectives. Ecosystem management is more relevant and acceptable to solving problems now than in the past because of new information from research and experience that make it possible to apply. Ecosystem management is a concept directed to examining the whole instead of individual parts. This "holistic" approach proposes that some entities, such as a unit of forest wildlife habitat, have a value independent of and greater than the sum of its parts. The concept is similar to the idea of holistic medicine, which is to treat the whole person instead of separate body organs.

4.2 ECOSYSTEMS

Ecosystem management considers a unit of land as a system, for example, a watershed. The parts of the system are the plants, animals (including insects), soil, climate, water, air, topography, and human influences comprising the whole. Any given factor in the system can be out of balance, which in turn affects other factors in a cycle of events. The cycle of events over time is the core of ecosystem management that must be understood; in most cases, the cycle of events transcends our own ecological longevity.

Ecosystem management is adaptive in character because our knowledge is limited and vague, and uncertainty is the normal result when trying to project future outcomes. Managers must make informed decisions about resource objectives, initiate the proposed actions, observe the outcomes of the actions, and then make changes that will meet the objectives. Ecosystem management in many respects is similar to the commitment of some organizations to total quality management. If a policy of ecosystem management is adopted, can it be practiced on the ground exactly as in concept? How does this relate to what the public wants from their forestland? There are three aspects of the problem that must be considered:

- Management and planning has to be modified to consider management units in larger areas than is being done now.
- The time perspective has to be directed to longer management periods.
- There must be a process of integrated resource management planning (IRMP) to develop alternatives for decision makers.

Most natural resource managers think of forest stands as small, which was acceptable and has worked well in the past to meet management objectives. But, these small stands, while creating tremendous diversity in "edge," may also be fragmenting the forest and reducing habitat for some animal species, such as those requiring interior forest conditions. Stands probably need to be delineated on the order of 50–100 ac within the boundaries of landscapes consisting of tens of thousands of acres. For ecosystem management, stands may have to be defined differently from a forest stand based on a combination of components on the ground.

4.2.1 Local and Formational

While ecosystems can be any size and shape, there are practical limitations to how small and how large an area can be managed as a unit. Natural ecosystems can be classified into five categories (Billings 1970):

1. Local ecosystems
2. Formational ecosystems
3. Continental or ocean-basin ecosystems
4. The earth ecosystem
5. The solar ecosystem

The local and formational ecosystems are of the most interest to managers. Local ecosystems are vegetation types or portions of types (patches) delineated by topographic features or other definable boundaries. An example would be an oak-hickory vegetation type in a watershed or a forest district. Formational ecosystems are large landscape areas such as a forest or part of an ecoregion in North America as defined by the Commission for Environmental Conservation (CEC 1997).

4.2.2 Landscapes

Large landscape areas are needed to get the mix of goods and services that cannot be provided in smaller units. Managing at the landscape level may mean the redefining of boundaries of forests and management units to conform to natural features such as watersheds instead of roads, streams, and fences. Most of the planning done today is based on 3–5 years of "here-and-now" time. This mind-set, however, is far too short when planning for a sustainable forest that requires the ecological time of succession from 60 to 250 years or longer.

4.2.3 Habitat Area

Habitat area results from the ideas of forest fragmentation and from the theory of island biogeography, which seeks to explain why the number of species on one island differs from that found on another. Darwin was the first to pursue this question in connection with his findings in the Galápagos Islands. MacArthur and Wilson (1967) integrated the concepts of ecology (including genetics) into the island theory of biogeography. Harris (1984) later translated the ideas of biogeography to isolated stands of old-growth (islands) surrounded by younger stands of trees (sea).

The idea of habitat area was also suggested by Elton (1966) in developing his image of the inverse pyramid of habitat; that is, tertiary animals, those carnivorous species at the top of the food chain, are always low in numbers for the area that they occupy relative to the large areas of habitat, such as for elk (order of magnitude [Om] 3.5, Figure 2.8), while small herbivores maintain populations of large numbers occupying small habitat areas (Om 0.5). Managers cannot discuss critical habitat or viable populations of species such as cougars and grizzly bears at the same scale as for squirrels and rabbits. Habitat area has recently received more attention due to large blocks of public forestland that are now scheduled for cutting or conversion to other uses and thereby will affect the habitat of some gamma species.

One of the major tenets of the theory of forest fragmentation is that the number of species in an island habitat is directly proportional to the size of the habitat area. This simply means that as the size of the island increases, the number of species increases. On the other hand, when larger areas of habitat are reduced in size, the number of animal species in the reduced habitat is proportionally reduced. The idea is not different from the species-area curve used by plant ecologists to determine plot size for sampling (Phillips 1959), but in the case of wildlife management, the plot size can be a block of land no smaller than is necessary to preserve a species.

The minimum size of forest blocks that must be maintained as cover is not yet well defined, but there is some indication that a reasonable area is between 75 and 100 ac for maximum species diversity for birds (Galli et al. 1976; Thomas et al. 1979c). An important consideration is that an increase in the number of species does not mean an increase in the density of species. Species of low-to-moderate mobility requiring several edge types to fully meet their food and cover requirements do not benefit from large blocks of old growth forest. However, as stated, some species need large blocks of mature and old growth because of the presence of specialized conditions in the interior of the forest stand (Galli et al. 1976). Wildlife managers must consider the size of the management unit together with forest habitat island size because the management unit is where the resources are provided for species to survive and reproduce.

The size of a management unit, which would include all native wildlife species, is determined primarily by the area needed to maintain a viable population of each species; alternatively, determination is by whether it is to include only selected species (featured species). To illustrate, one cougar, a landscape species, may require a minimum of 74,100 ac for a home range (Table 2.13). If 74,100 ac are all that are available in a forest, cougars may not be able to survive and reproduce. The area is too small to maintain a viable population. Thus, it is evident that more than the home range of a single individual needs to be considered. By contrast, it is not realistic to manage units in the 5-ac size necessary for a low-mobility species unless it is on the federal list for threatened and endangered species. A cougar (Om 4.5, Figure 2.9) confined to the home range of a rabbit (Om 1.0) could not survive, but several thousand rabbits can exist within the home range of a cougar.

The advantages of maintaining a small unit (e.g., 500 ac) may be outweighed by the cost of administration, except in the case of habitat for an endangered species.

Conversely, planning for and maintaining wildlife habitat in areas from 30,000 to 50,000 ac may not be possible because it is difficult to provide information with the detail required to maintain viable populations of all native species. However, habitat management guidelines for elk, an economically important species in Montana and Idaho, suggest that landscape units from 30,000 to 150,000 ac may be necessary to take into account the size of adjacent project-level effects that are from 3,000 to 10,000 ac (Christensen et al. 1993). One solution to the problem of management unit size is to select a size that includes the home range of a majority of species in the forest and then combine management units by providing connecting corridors for species of high mobility. A management unit in the range of 7,000 to 10,000 ac seems to be a reasonable compromise as it includes a large number of the species inhabiting most forest types.

4.3 WILDLIFE HABITAT RELATIONSHIPS

In recent times, biologists have had to devise new ways to manage wildlife populations and habitats because legislation has introduced terms and concepts, such as diversity and management indicator species (MIS), into the planning process. As a result, several new approaches to management are being tried by resource agencies, and this experimentation will continue until there is some agreement on which one best meets the various legislative intents while maintaining management integrity. Because of the diversity of many conceptual approaches to a variety of systems, confusion often arises about which concepts to adopt or which techniques to use to formulate and meet the requirements of a given plan.

Many administrators are bothered by the fact that resource agencies do not use standard techniques in the decision-making process across an array of land use plans (see Chapter 1 on the principle of complementarity). There can be several correct answers to managing wildlife populations and habitats, so the manager is free to select from a variety of approaches to accomplish a set of objectives. The choice depends on

- The degree of complexity in the data needed to make sound management decisions
- The amount of data available
- The amount of money needed to meet the objectives of the management plan

The realization that managers have several systems from which to choose provides them with a richer set of opportunities to devise a successful plan and more alternatives by which to explain decisions. In general, the current systems of managing wildlife are either single species or multispecies oriented. However, there is now a trend to integrate these into one system based on management of ecosystems. In the following sections, we will explore approaches to management with consideration for what will be possible in the future as new and more precise information becomes available.

Wildlife habitat relationships (WHRs) are an "eco" systems approach to comprehend all the associations with and uses of habitat by wildlife. A *system* is a set or arrangement of related or connected things forming a unity or whole. The *eco* part of ecosystem directs attention to the fact that the systems of concern are ecological and natural. The use of WHR is a way of thinking about the complexities of plant-animal interactions in an ecosystem or executing a complex set of instructions in an orderly way to achieve a stated objective, as in a monitoring scheme.

Information from both single- and multispecies management can be integrated in a relationships system to meet the objectives of managing landscapes on a sustainable basis. The process depends on identifying all linkages (relationships) and interactions between plant and animal species in a given ecosystem. The WHR as a process resulted from legislation requiring land management agencies to develop management plans with goals, objectives, and alternatives along with a monitoring system to track progress. The ultimate goal of the WHR is to ensure that wildlife is considered in implementing ecosystem management. The WHR can either be applied in general to a way of carrying on the work of a resource management agency or in a strictly ecological way as a means of modeling plant and animal interactions. The first approach was adopted as policy by the U.S. Forest Service (Nelson and Salwasser 1982). Other land management agencies are using the same approach but have modified procedures to meet their own set of governing state or federal laws and regulations.

The ecological approach to using a WHR in a systematic way has been documented by Patton (1978), Thomas (1979a), Verner and Boss (1980), Hoover and Wills (1984), Morrison et al. (1998), and DeSteffano and Haight (2002), but the study of the association of animals with their habitat has been in progress for many years. Attempts at a functional classification was in progress in Texas as early as the 1940s, resulting in a scheme classifying 49 mammals and 171 birds into nine categories from "conspicuously insectivorous birds" to "tree protectors" (Taylor 1940). The intent of the classification was to indicate how nongame species habitat was ecologically interrelated to the habitat of game animals.

A modification of Taylor's efforts resulted in a classification for North American mammals and birds by forest habitat preference (Yeager 1961). In this scheme, birds and mammals were placed into three groups according to whether their habitat was primarily forest or brushland, whether they associated with forest or brushland, and whether they only rarely, if ever, used woody cover. The significance of this form of synthesis is greater today than it was in 1961, and Yeager's summary is still appropriate:

> The ecologist capable of arranging all species of mammals and birds, or all vertebrates or animals, into a habitat classification fully considerate of vegetation, terrain, water, elevation, exposure, latitude, and climatic factors would indeed provide a vast fund of useful knowledge. (Yeager 1961, p. 671)

Because of the amount of data that has accumulated since the 1940s, present-day wildlife managers understand habitat relationships in more detail and are better able to develop a systematic scheme for using plant and animal data. The relationships hierarchy, when included in a relational database, can combine the multispecies management systems of guilds, life forms, habitat profiles, and ecological factors into a detailed but easy-to-use system in which information can be obtained for the level needed to meet management objectives or for decision making. The different levels of abstraction allow "nested questions" to be asked about the relationships.

Many useful techniques have been developed to assist field biologists in managing wildlife populations and habitats. In some cases, as for basic habitat models or scorecards, hand manipulation of data is adequate and is still a common procedure. Some of the visual charts, graphs, and figures that have been developed over the years are effective in presenting the flow of information in plant-animal relationships. These types of manual and visual tools are also preliminary forms suitable for integrating into sophisticated computer simulation models. However, when hand manipulation of data is no longer practical, managers must then use computers to increase their efficiency and to manage the massiveness, detail, and complexity of the data that have been developed.

Computers have, for the first time, given resource managers the capability to sort through and organize millions of pieces of data quickly to provide comparisons, mathematical analyses, and statistical summaries. While the computer clearly has many advantages over hand manipulation of data, the user must always be aware that computers are controlled and programmed by humans who also enter the data. Computer output is only as good as the quality of the stored information. The rapid development of microcomputers has brought about a significant change in the way wildlife habitat and species information is gathered, recorded, and used in decision making.

The period from 1980 to the present has been one of explosive development not only in increasingly powerful computing devices but increasingly complex software to store and retrieve information. These improvements have led to better and more efficient ways of analyzing the data that has been accumulating in books, periodicals, and storage files for years. As a result, managers are now developing larger databases and experimenting with increasingly sophisticated and realistic habitat simulation models to use in solving management problems.

Models reduce the ambiguity and complexity inherent in understanding the involved and intricate network of relationships found in every ecosystem. Both manual and computer techniques directed at understanding and producing management plans have a place in resource management. In this chapter, examples of both kinds of techniques are presented. The purpose is to suggest ways in which various techniques can be used singly or, more often, combined into powerful habitat relationship systems to assist managers in decision making.

4.3.1 SINGLE SPECIES

For years, game and fish departments have been using single species management for hunted or fished species. This approach has resulted in many published federal aid documents on management of squirrels, cottontails, deer, elk, quail, and waterfowl. Excellent references are available that document the requirements of game animals such as the *Black-Tailed Deer of the Chaparral* (Taber and Dasmann 1958), *The Wild Turkey and Its Management* (Hewitt 1967), *The Pronghorn Antelope and Its Management* (Einarsen 1948), and the books on deer (Wallmo 1981) and elk (Thomas and Toweill 1982). These are only a few of the hundreds of publications contributing to the understanding of a single species life history and habitat management.

In the early years of the wildlife profession, information on a single species tended to be at a broad scale concerned with the distribution and general food and

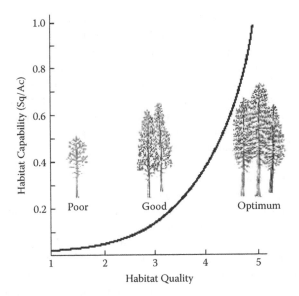

FIGURE 4.1 Habitat capability model for the tassel-eared squirrel in Arizona. (From Patton, D. R. *Wildl. Soc. Bull.* 12:409–414, 1984.)

cover requirements. As information accumulated, data became more refined and specific to use in developing written habitat guidelines that conform to legal requirements of legislation. The data are often translated into a rating system for habitat, as has been done for big game in Montana (Cole 1958), California (Dasmann 1948), and New Mexico (New Mexico Game and Fish Department 1973). These guidelines have proved to be a real asset to field biologists in making management decisions. However, their use has led to recognition of the need to quantify requirements to better predict the effects of land treatments on wildlife habitat over time.

The simulations of forage for elk in Montana (Giles and Snyder 1970) and for deer in Colorado (Wallmo et al. 1977) are examples of techniques used for quantifying data and extending its utility for management. These investigators used forage as the critical component of habitat; research on other species has indicated that food and cover variations can successfully be integrated to develop habitat capability models (HCMs) or habitat suitability indexes (HSIs). Both model types use an integrative approach incorporating several habitat variables to arrive at a composite rating. For the tassel-eared squirrel, the habitat quality rating is based on tree density, size, and grouping to provide information on the habitat requirements for a given number of squirrels (Figure 4.1).

4.3.2 MULTISPECIES

The concept of *multispecies management*, while not new, received little attention until federal legislation required consideration of all species and is now an accepted approach in land use plans. Multispecies management involves *grouping species with common habitat requirements* and then managing for identified, overlapping, or critical habitat components. A good example of multispecies management is the management of snags for cavity-nesting birds (Scott and Patton 1989). Another is managing for 14 prey species to provide a food base for the northern goshawk (Reynolds et al. 1991). The northern goshawk is a habitat generalist that uses a variety of successional stages and stand structure because these conditions provide snags, downed logs, woody debris, and large trees that are habitats for food sources such as squirrels, chipmunks, woodpeckers, jays, and rabbits. In addition, the goshawk habitat contains small-to-medium openings and patches of dense midsize trees along with 50–60% of the forest in mature and old forest condition. This type of mixture of stand, tree sizes, and tree density also will be habitat for other species, such as deer, elk, turkey, and band-tailed pigeon.

The concept of multispecies management does not rule out a single species emphasis, such as for the goshawk, but it does mean that management for any single species should include the effects of this management on other species in the defined ecosystem (Patton 1993). The ultimate goal is a diversity of habitat in various sizes and stages in ecosystems (vegetation types) across landscapes. In the final analysis, multispecies management is fostering the change from an emphasis on species to a focus on habitats with the objective of maintaining healthy, diverse, and sustainable ecosystems.

Multispecies management techniques are presented in the form of guilds, life forms, and ecological factors. Managers using these various approaches are all trying to achieve the same goal, which is to manage the forest as a *commons* for many species. This is an appealing approach because it reduces the cost of providing habitats for many single species by manipulating habitat through coordination measures with other uses of timber, range, watershed, and so forth.

4.3.3 FEATURED SPECIES

A modification of single species management is *featured species* management by which one animal is selected for an area and management efforts are concentrated to fulfill its needs through coordination of timber and wildlife habitat activities (Holbrook 1974). Small areas within the habitat of the featured species are selected to meet the needs of other species (Figure 4.2). If properly planned and implemented, featured species management provides a variety of habitats for other species that overlap the requirements of the featured species. For many years, featured species management has been used successfully by state agencies primarily for game animals such as deer, elk, and turkey.

The data used in single species models and guidelines provide the basis for future multispecies habitat models, but at present such models are only in the conceptual and early developmental stages. Progress has been made in developing simulation techniques that identify single species as a component of a main program. For example, the ECOSIM (ecological simulation) model developed for ponderosa pine in Arizona includes information and landscape-level projections for the tassel-eared squirrel as well as for deer and elk forage (Rogers et al. 1981). There is little doubt that single species management is currently the leading approach and will continue to be important in the management of game animals and wildlife classified as threatened and

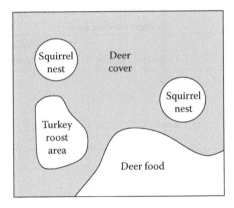

FIGURE 4.2   Deer is the featured species, but habitat is provided for other species as well.

endangered. However, research on food webs and food chains provides evidence that all animals facilitate nutrient cycling and energy flow, and that long-term ecosystem health depends on diverse animal species.

4.3.4   GUILDS

Guilds are defined as groups of species exploiting the same class of environmental resources in a similar way (Root 1967). A guild groups species for habitat reasons without regard to taxonomic criteria. The important consideration when developing guilds or related systems of species grouping is whether the result provides information that can be used in decision making. Classifying animals into carnivores, herbivores, gleaners, and peckers, for example, does not provide adequate detail for management purposes. Such classifications, however, do provide a broad hierarchical level as a base for developing models and relationships.

Typically, guilds are contained in a matrix table, with one axis representing the feeding habitat and the other axis the reproducing habitat. In the past 30 years, much useful research on the theory of guilds with an emphasis on birds has been completed (Severinghaus 1981; Short and Burnham 1982; Mannan et al. 1984; Verner 1984). A good example of an easy-to-understand guild matrix for birds was developed for the pine-oak woodlands of California (Figure 4.3) (Verner 1984). The primary feeding and nesting zones form the two axes of the matrix. Within the matrix, birds are assigned to each cell according to their feeding and nesting habits. The matrix has a potential of 30 guilds if all cells contain information.

4.3.5   LIFE FORM

Thomas et al. (1979b) have successfully used a life form approach of the multispecies concept of management in the Blue Mountains of Oregon, and the concept has been extended to other areas (Brown 1985). A *life form* is a single category synthesizing information on where an animal feeds and breeds (Haapanen 1965). All the vertebrates (26 amphibians and reptiles, 263 birds, and 90 mammals) known to occur in

Primary Feeding Zone

Primary Nesting Zone	Ground	Shrubs	Tree Boles	Tree Canopy	Snags
Ground	tuvu caqu cora weme rcsp				
Shrubs	road bhco lego brto	wren cath			
Tree bole	mke scow cofl star	howr bewr	nuwo wbnu	acwo atfl alti webl	vgsw
Canopy	rtha modo ghow ledw scja hofi	bush		coha anhu weki bggn phal huvl noor	
Snags					
Breeds elsewhere	amro rsto gcsp	heth rcki wcsp fosp lisp	ybsa	ssha btpl yrwa deju	

FIGURE 4.3 Guild matrix for birds using the pine-oak woodlands in California. (Birds are identified by using a four-letter code derived from their common or scientific name.) (From Verner, J. *Envir. Manage.* 8:1–14, 1984).

the Blue Mountains were ultimately grouped into 16 life forms. The most obvious advantage of both the guild and life form concepts is that a large amount of information can be condensed into tabular form for quick reference. The format of the life form system has been implemented in other states, such as Colorado, and is still in use for many national forests.

4.4 ECOLOGICAL FACTORS

Management plans must clearly provide an array of ecological factors that have to be evaluated, manipulated, and monitored to realize the wildlife objectives set by a plan.

Several different concepts for understanding and providing for ecological factors have been developed. Guilds and life forms are direct ways to group species with overlapping physical habitat requirements. There are, however, indirect ways to approach multispecies management through the notion of ecological factors, which are categorized as indicators, diversity (biodiversity), edge effect, and habitat relationships.

4.4.1 INDICATORS

Indicators were first used in resource management by plant ecologists working with soil productivity and agricultural crops (Schantz 1911). In the 1930s, vegetation condition and trend techniques were developed for range management using plant indicators (Talbott 1937). Modified forms of these techniques are still in use today. The term *management indicator species* (MIS) was introduced to the wildlife profession by the National Forest Management Act (NFMA) (U.S. Congress 1976). This type of indicator is different from those described in ecological literature.

An *ecological indicator* is a plant or animal so closely associated with a particular environmental factor that the presence of the species is clearly indicative of the presence of the factor (Clements 1916, 1949; Patton 1987). In most cases, a carefully selected indicator identifies a *cause-effect relationship*; although the cause cannot always be easily detected, identified, or quantified, the effect is clearly desirable or undesirable. Every plant or animal is a product and measure of its environment and therefore an indicator of some habitat factor because of its inherent need for that factor.

Ecological theory and experience suggest some general criteria for selecting indicators. First, indicators should have a narrow and specific tolerance to the factor being indicated, so that by definition the relationship has a high degree of certainty. Species with wide ecological amplitudes, which can use a combination of factors, none of which is limiting and many of which can be substituted one for another, are not suitable indicators.

Second, in most cases plants as habitat components are the best indicators of change since they are directly influenced by environmental factors. Plants are nonmobile and easy to count and indicate change with a high degree of certainty. Finally, because single plants or a particular set of species may indicate the successional stages of plant communities, the community as a whole is the best indicator of the environment. Whether this community approach is equally valid for groups of animals has not been substantiated. Ecological literature contains a large number of examples of the use of indicators, so it is not surprising that the framers of environmental resource legislation extended the concept to monitor progress in land management plans. Regulations implementing the NFMA defined *management indicator species* as

1. Endangered and threatened plant and animal species identified on state and federal lists for the planning area
2. Species with special habitat needs that may be influenced significantly by planned management programs
3. Species commonly hunted, fished, or trapped

4. Nongame species of special interest
5. Plant or animal species selected because changes in their population are believed to indicate effects of management activities on other species of a major biological community or on water quality

The five categories provide great latitude in what can be selected as an indicator. Category 5 suggests that an indicator species can be used to monitor the effects of management activities on the population trend of a group of species, but research showed that this may not be a valid use of indicators (Mannan et al. 1984; Verner et al. 1984; Szaro 1985). The important consideration for using any indicator is to decide *what is to be indicated* and then determine if there are cause-effect relationships that can be indicated with a high degree of certainty.

4.4.2 BIOLOGICAL DIVERSITY

A term receiving much attention from forest biologists is *biological diversity* (also termed *biodiversity* or simply *diversity*), but the topic is not new to the wildlife profession, as detailed in Pimlot's (1969) review. Diversity was defined and discussed in Chapter 1, and one should not forget that understanding diversity is one part of understanding ecosystem functioning. Strategies for maintaining biological diversity include horizontal and vertical tree components, the layering effect of these components, and the numbers and distribution of plant and animal species in a given area. Geometric design and space are also included in strategies to maintain diversity, and one component of the design factor is the effect of edge.

4.4.3 EDGE

The geometric figure possessing the greatest area and least perimeter or edge is a circle (Figure 4.4A). Edge can be quantified by relating it to area (Patton 1975). If the ratio of circumference to area of a circle is given a value of 1, a formula can be derived to compute an index for any area to compare with a circle. A value larger than 1 is a measure of irregularity and can be used as a diversity index (DI). For a 1-ac circle, the circumference C is 740 ft (rounded) with an area A of 43,560 ft^2. The formula to set the ratio equal to 1 is

$$1 = C \div 2\sqrt{(A \times \pi)}$$

This same formula is often used by limnologists to express shoreline irregularity of a lake. The next step is to restate the formula for habitat diversity as an index:

$$DI = TP \div 2\sqrt{(A \times \pi)}$$

where TP is the total perimeter around the area plus any linear edge within the area. A 1-ac square is 208.7 ft on a side; after multiplying by 4 and rounding up the answer, the perimeter is 835 ft (Figure 4.4B). Solving for these values,

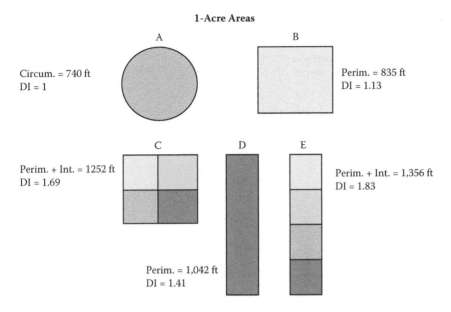

1-Acre Areas

Circum. = 740 ft
DI = 1

Perim. = 835 ft
DI = 1.13

Perim. + Int. = 1252 ft
DI = 1.69

Perim. + Int. = 1,356 ft
DI = 1.83

Perim. = 1,042 ft
DI = 1.41

FIGURE 4.4 Comparison of edges created by different patterns. (Circum., circumference; Perim., perimeter; DI, diversity index; Int., interior.)

$$DI = 835 \div 2\sqrt{(43,560 \times 3.1416)} = 1.13$$

By rewriting the formula, DI can be expressed as a percentage, where

$$\% = (DI - 1) \times 100$$

Thus, the computation for a 1-ac square with a DI of 1.13 is $(1.13 - 1) \times 100 = 13\%$.

This figure simply means that a 1-ac square has 13% more perimeter than a 1-ac circle. A square mile by analogy also has 13% more perimeter than a circle of the same area. The formula is valid regardless of whether the units are in the English units of feet and acres or in hectares and meters. Dividing the 1-ac block of the square example into four units of different vegetation types increases the DI to 1.69 (Figure 4.4C). A block of the same size but long and narrow has a DI of 1.41 (Figure 4.4D). If the long and narrow 1-ac block is divided into four smaller units, then the DI is increased to 1.83 (Figure 4.4E), and the TP is computed by adding the outside plus the perimeter of the three inside edges.

The DI assumes that the perimeter of the area is all edge; however, in application this will probably not be the case except for natural or created openings. In other situations, the area could be a boundary (such as an arbitrary unit dividing line) where part of the perimeter is not edge. In this case, that part of the boundary is not included in the determination of TP. The point to keep in mind is that perfect circles and squares do not occur in nature, but they present a base or standard for comparison. In addition, where two vegetation types meet, any deviation from a straight boundary to one with curves is also a degree of diversity.

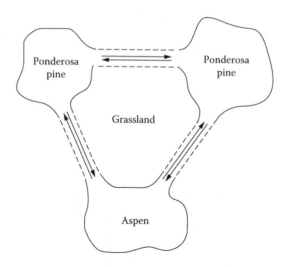

FIGURE 4.5   Corridors provide cover for species moving between habitat islands.

4.4.4   HABITAT CORRIDORS

In many regions, large blocks of old-growth timber have been removed, so the problem now is one of providing corridors so wildlife species can move between habitat islands (Figure 4.5). This has been termed the "island archipelago approach" (Harris 1984). Evidence for the use of corridors has been developing since the 1970s; thus, MacClintock et al. (1977) found that small areas connected by corridors to larger areas supported a species composition comparable to that of the larger area.

One system to maintain old-growth forest conditions is by maintaining habitat islands in *long rotations* surrounded by normal rotation patches (Harris 1984). The key to the system is scheduling stands to be harvested to maintain the island and corridor effect necessary for species diversity. In the process, care must be taken not to increase the number and size of the habitat islands in a combination that will in turn lead to homogeneity over a large area. For example, a large number of small openings of the same size when increased in number reduces the size of the remaining forest stands so that both openings and stands are the same size. In such circumstances, homogeneity is created by the evenness of the spacing of stands and openings.

In a study of cougars in the Santa Ana Mountains in California, where habitat was fragmented and urbanized, corridors of 0.92 to 3.7 mi were used by juveniles to disperse to new home ranges (Beier 1993). Corridor users can be grouped into "passage species" and "corridor dwellers" (Beier and Loe 1992). Passage species, such as large herbivores and carnivores, use corridors to pass between two areas in a brief time and for a specific event (dispersal of young, migration, etc.). Corridor dwellers (small mammals, reptiles, etc.) take a long time to pass through the corridor.

In the southwestern United States, riparian vegetation forms corridors traversing habitat from high-elevation spruce-fir and ponderosa pine forests to the lower pinyon-juniper woodlands. In addition, stringers of ponderosa pine and pinyon-juniper, resulting from differences in soils and exposure (inherent edge), penetrate the surrounding vegetation, increasing not only the edge but also the food and cover areas. Corridors of these types exist in many regions and must be protected and maintained as critical wildlife habitat (Thomas et al. 1979c).

4.4.5 GAP ANALYSIS

In the early years of the wildlife profession, considerable time and effort were directed to drawing food, cover, and water locations on acetate overlays, as I did for the Santa Fe National Forest in 1963. These acetate overlays were an early approximation of what was to come with the development of geographic information systems. When resource overlays were placed one on the other on a base map, it was apparent where there were gaps in the juxtaposition and dispersion of the resources necessary for animals to survive and reproduce. One of the most common ways to determine where to provide water for habitat improvement was to delineate zones of influence at given distances from the water source. The intent was to provide a water source within travel distances of 1–5 mi for use by deer, elk, and livestock. This type of gap analysis identified the areas that had priority for habitat improvement given that cover was available.

The early form of gap analysis using overlays has given way to more exact techniques using systems that are capable of storing and retrieving large numbers of resource overlays on one map. This type of analysis can be used to determine the distribution and status of several animal species with different land management and land ownership patterns (Scott et al. 1993). In the process of identifying gaps, the extent of habitat fragmentation can be determined, and this in turn will reflect the need for habitat corridors. Gap analysis can be done for any scale of resolution using information from 1:15,840 maps to high-altitude photographs and multispectral space images.

4.5 HABITAT MODELS

A model is an intellectual representation of a real-world situation (Figure 4.6). In developing a model, the modeler must identify a representation (abstraction) that accurately and comprehensively defines the problem. Once a model has been developed, the results must constantly be interpreted, and decisions must be made whether more data are needed or a modification of the model is required.

Humans use models every day to make decisions because they are constantly drawing on experience and learned patterns of using data stored in their memory. As experiences and patterns of thinking accumulate and are in turn used to make further decisions, we are testing or retesting a consciously or subconsciously formed model. If our decision is correct, our mental model is validated and will be used again in similar circumstances. If our decision proves unsound, we revise the model in light of this additional experience and new data and then try again. In effect, we routinely build models in our mind and repeatedly test them by making decisions and assessing the outcomes of these decisions.

We build models because they help us categorize and organize data and thoughts and communicate ideas. Conceptual Equation 1.1 (Section 1.4.1) is a model

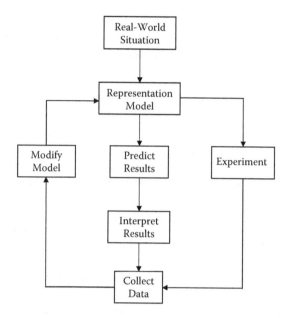

FIGURE 4.6 Models are a representation of a real-world situation.

categorizing and organizing all the direct factors affecting the ability of an animal to survive and reproduce. As a conceptual equation, it is a useful tool to synthesize an immense body of data and to communicate ideas. Models can be used to make predictions, and frequently they can be used to compare management alternatives.

In wildlife and natural resource management, decisions must be made even though there is often a lack of good data on the use of habitat by a particular wildlife species. Unfortunately, this is the rule rather than an exception. But, even when data are lacking and the biological processes are not well understood, techniques have been developed to improve decision making under conditions of uncertainty.

4.5.1 FACTORS TO CONSIDER

Models do not just happen. They have to be deliberately planned to solve a problem; there are factors to consider. Since models are a product of human thinking, they must be developed so that others can use and understand what the model does. Because a transfer of knowledge takes place between the model and user, the model must have a structure that facilitates the transfer. Models are made up of an input function, a manipulative function or algorithm, and an output function (Figure 4.7A). Users need not understand the algorithm that manipulates the data, but they must be able to understand the structure of both the input and the output information and how the latter can be used in decision making. Although the user may not understand the internal workings of a particular model, he or she cannot be relieved of the responsibility of assessing the validity of the results.

Models use different levels of *resolution*, which refers to the programmed ability to separate components (Starfield and Bleloch 1986). For example, the results of one

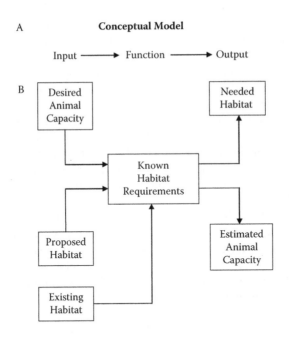

A   **Conceptual Model**

Input ⟶ Function ⟶ Output

FIGURE 4.7 Models require an input, a manipulative function, and an output.

model may address total herbage production, while those of another separate production into woody plants, forbs, and grasses. If resolution is too coarse, as may be the case with total production in the first example, the model is generally inadequate. If the resolution is too fine, too much time is spent entering or manipulating irrelevant detail. Resolution also has to take into account the level or scale of decision making. For example, models designed to provide answers on a regional scale are invariably too coarse for local use.

Models also vary in the degree of *detail* incorporated. In estimating herbage, should the user record quantities to the nearest ounce (gram) or to the nearest pound (kilogram)? The question is, Will the decision to be made be affected by using the coarser herbage quantity? The impact of the effects of detail can often be determined by conducting a sensitivity analysis. A *sensitivity analysis* is a means of systematically determining how changes in a variable or parameter affect the output of the model. If a large change is needed in an input variable for a small change in the output, the variable being treated is not very sensitive. However, if a small variation in input results in a large change in the output, the variable may be too sensitive for practical field use.

The modeler must constantly think of how the output information is to be used in decision making and how much effort is required to collect data for input. The variables to be included or excluded define the scope of the model. If three variables will provide a 95% confidence interval while two variables will provide a 90% confidence interval in the output, is the inclusion of the third variable justified? Another important factor in model development is time. Is the model to be used for present time (3–5 years), ecologic time dealing with plant succession, or short time, such as in population equations? In model building, most of the questions relating to scope, time, sensitivity, detail, and resolution are automatically taken into account by knowledgeable programmers.

Models are generally of two types: conceptual or empirical. *Conceptual* models represent thoughts and ideas in a qualitative and deductive way rather than in terms of numbers. Input, program function, and output can be formulated in a detailed and recognizable way to indicate how a conceptual model operates (Figure 4.7B). While the arrows show a one-way flow of information, they can be reversed so that biologists seeking to maintain the population level of a certain species can determine the amount of habitat needed, whether it currently exists or is proposed. Conceptual models generally precede the more sophisticated *empirical* models, which use both qualitative and quantitative data.

Empirical models are based on the ideas and relationships contained in conceptual models but augmented by numeric data to provide a systematic means of making a decision. Quantitative empirical models are often *deterministic*. They predict a future event by using measurable sets of factors and relationships. For example, if the rate of increase r in a population of animals is known to be 0.02 and the base population is 1,000, then the increase I in the next year is

$$I = r \times P = 0.02 \times 1,000 = 20$$

Present-day wildlife biologists can only hope that at some time in their career they will possess this type of data, and that all their problems will be as simple to solve as the one in the example. Increasingly, models are being developed to use *stochastic* phenomena. These types of models are designed to include the statistical probabilities that some special event will or will not happen. Stochastic models can operate at a level of uncertainty, for example, as required in pattern recognition (Williams et al. 1977).

4.5.2 DECISION MAKING

In the past several years, there have been adequate warnings given to managers and developers about the proper use of models in making decisions (Verner et al. 1984), and there have been suggestions presented to improve the model development process (Jeffers 1980). Before undertaking a major project to develop a model for use by managers, it is incumbent on the developer to seek help and advice from those with experience and background not only in the process itself but also in the subject area of the model. The best situation is to have several people with varied backgrounds working as a team (Bunnell 1989), as is often done in database development (Patton 1978), or an expert system using a Delphi technique (Crance 1987; Mollohan 1988).

In developing models, do not overlook biologists with many years of experience. These people may never have developed a model on paper, but every day they are in the field they build models in their mind and use them to make decisions. Trying to duplicate this process for a model requires developers to put themselves in the manager's position to understand the kinds of information needed to make an informed decision. Because of the large number of laws and regulations requiring many different types of data, managers now are in critical need of information to make decisions. If researchers and managers do not want their information used, then it should not be written—once on paper, it will get used even if it is just a wild guess.

4.5.2.1 Decision Tree

Another approach to decision making is the use of *decision tree* and *Delphi* techniques (Crance 1987) or an expert system. Because of the recurrent need to make many decisions in short time periods with inadequate data, such as in national forest plans, a trend has developed to use techniques that produce results more quickly and less expensively than models that depend heavily on experimentation or the collection of field data.

The intent of a decision tree model is to chart and document a course of management action based on a logical assessment of a set of alternative rules. Pairs of empirical propositions are presented, and one is then selected over another. This process of selection is repeated until a decision is reached; both the format used and the logical steps taken are analogous to those of a dichotomous key for determining a plant or animal species. Decision trees are often depicted on a chart showing the pathways that can be taken to a decision. These displays help the user to visualize the total set of alternatives and provide some confidence that what the user has selected is the right decision.

The Delphi technique has had considerable use in the social and behavioral sciences but in recent years has become more popular in solving natural resource problems. The Delphi technique is a process similar to the expert opinion technique except that it is more detailed and time consuming. Members of the group using a Delphi technique may not know each other; the process of focusing on the selection of alternatives and their outcomes is generally conducted through the mail or by electronic means. In addition, some statistical analysis can be done in the Delphi analytical process.

4.5.2.2 Expert Systems

Decision making at the local level can often be enhanced by developing a model for a specific area or project to display alternative treatment effects. Models for decision making need not always be in a form suitable for computer use. The initial step in developing a habitat model involves defining the requirements by kinds, amounts, and distribution of food and cover for a species or group of species. The second step is to quantify these habitat components and group them into quality classes (poor to excellent) to rate a habitat for a given species or group of species.

One approach is to rate habitat for individuals or groups of species on a scale from 1 to 5. The rating is based on the opinion of a group of experts or sometimes even on one experienced person's opinion. The rating by the experts or manager is based on the knowledge of the food and cover requirements of the species involved and how these requirements are met by the existing habitat or will be met as the habitat is changed by a proposed timber treatment. Rating of habitat can be done on any scale (project area, planning unit, watershed, and district), depending on need.

Table 4.1 presents a rating for an imaginary district with four planning units. The scores and ratings for the example district are based on four units for six species.

TABLE 4.1

Wildlife Species Ranked by Quality of Habitat on a Four-Unit District

Wildlife	Unit Score				District	Rating
Species	A	B	C	D	Score	(%)
Deer	3	3	2	1	9 (9/66)	45
Elk	2	4	5	2	13	65
Turkey	1	3	2	3	9	45
Squirrel	5	4	1	4	14	70
Cottontail	4	3	1	2	10	50
Grouse	2	5	3	1	11	55
Total	17	22	14	13	66 (66/120)	55

Note: Scores are based on a scale of 1–5, 5 is optimum. The maximum rating possible is 120 (6 species × optimum score of 5 × 4 units).

Each unit is given a score for habitat quality by species, where 0 = none or eliminated, 1 = poor, and 5 = optimum. Species rated for each unit vary due to differences in vegetation types and successional stages supporting differing amounts and types of food and cover. In the example, the total score is 66. Column totals give scores for each unit, while row totals show the score for each species. The base value of the district is then calculated to compare actual habitat value to an ideal or optimum habitat. If each habitat were given a rating of 5 (excellent), then 6 species × 5 maximum ratings × 4 units equals 120. The district base is ideally 120, but the present quality is 66; this can be expressed as a rating of 55% (66 ÷ 120). The assumption underlying the rating is that the lower the percentile is, the fewer the animals the unit on a district can support. Also, the more closely the rating approaches 5, the greater the number of animals a unit on the district can support.

Additional species can be added to the model to more adequately reflect the total wildlife population native to the district. By so doing, the base (denominator) will be increased, but the quality rating will always remain between 1% and 100%. A scale of this type easily lends itself to setting goals and levels of management intensity (Table 4.2). The district goal might be to raise deer habitat quality from 45 to 55%. Alternatively, improvement of habitat might be undertaken for all species having similar requirements or only featured species to bring the habitat quality to any level consistent with the financial resources available.

It is wise to record on a form the information used to arrive at these manually maintained ratings. References are critical if habitat ratings are to be included in management plans or environmental impact statements since natural resource management plans, with increasing frequency, go through litigation. The basic assumption for the local habitat model is that habitat quality over the long run determines population density of the targeted species.

TABLE 4.2

Dollar Values Required to Achieve a Management Goal for a Particular Species

Estimated Habitat Quality			Improvements	
Species	Present	Desired	Needed	Cost
Deer	45%	55%	20 openings, 10 ac each	$40,000
			Water developments	$10,000
			Aspen regeneration	$35,000
			Seeding grasses and forbs	$50,000
			Total	$135,000

4.5.3   Pattern Recognition

Pattern recognition for wildlife management was developed as a direct response to the needs of wildlife managers, who must often make decisions in an environment of considerable uncertainty. Such methodologies are becoming increasingly useful as biologists become more familiar with this format and the ease with which field data can be collected and structured into a model (Grubb 1988).

An example shows how the pattern recognition procedure is used. Consider a hypothetical turkey habitat with 10 units occupied and 10 units unoccupied. For the units to be occupied, some attributes necessary for perpetuating the population must exist. By selecting five attributes from the occupied sites that previously have been identified as contributing to turkey habitat, a model can be developed. These same attributes may or may not occur in 10 units of habitat that are unoccupied. Using presence-absence information, a set of probabilities is computed for each attribute (Table 4.3). For Attribute 1, which occurs on seven occupied sites, the probability is $7 \div 10 = 0.70$, and for the unoccupied sites it is $3 \div 10 = 0.30$. The same computation is used for the other attributes for occupied and unoccupied sites.

A new unit of habitat is to be evaluated for the probability of being occupied by turkeys. In the new unit, the following information is obtained to be compared with the occupied and unoccupied information:

Attribute	Present
1	Yes (+)
2	Yes (+)
3	No (−)
4	No (−)
5	Yes (+)

A set of probabilities is computed using Bayes's theorem (Mason 1970). The computations for the habitat to be evaluated are as follows for the model of occupied sites:

$$P_O = (0.70)(0.60)(1 - 0.80)(1 - 0.90)(0.40) = 0.00336$$

TABLE 4.3

Data from 20 Sample Units with Turkey Habitat Attributes

Habitat Attributes	Occupied (10 units) %	Unoccupied (10 units) %
Five openings from 10 to 20 ac (4–8 ha)	7 (.70)	3 (.30)
Two trees/ac > 20-in. (50-cm) dbh	6 (.60)	2 (.20)
Average trees > 9-in. (23-cm) dbh	8 (.80)	4 (.40)
Oak > 10-in. (25-cm) dbh is present	9 (.90)	5 (.50)
Herbage production in openings > 1,000 lb/ac (1,120 km/ha)	4 (.40)	7 (.70)

Attribute 1 and 2 are derived directly from Table 4.3, but Attribute 3 does not occur on the site to be evaluated (as indicated by the minus sign), so the computation is $1 - 0.80$, with 0.80 coming from the frequency data for occupied habitat for Attribute 3. The same procedure is followed for all items with the minus signs in occupied and unoccupied sites. Next, the model of unoccupied sites is calculated:

$$P_U = (0.30)(0.20)(1 - 0.40)(1 - 0.50)(0.70) = 0.0126$$

With these calculations, the probability of the unknown site being inhabited can be calculated by the following formula:

$$P_H = (P_{PO} \times P_O) \div [(P_{PO} \times P_O) + (P_{PU} \times P_U)]$$

where P_H is the probability of evaluated habitat being turkey habitat, P_O is the probability of attributes being in occupied habitat, P_U is the probability of attributes being in unoccupied habitat, P_{PO} is the prior probability of occupied habitat, and P_{PU} is the prior probability of unoccupied habitat.

Since the chances that the site to be evaluated is turkey habitat are unknown (P_H), a prior probability (best guess), such as a 50:50 chance that it will be occupied (P_{PO}) or unoccupied (P_{PU}), is assigned; thus, $P_O = 0.5$ and $P_U = 0.5$. With this estimate, the calculated probability that the site is turkey habitat is

$$P_H = (0.5 \times 0.00336) \div [(0.5 \times 0.00336) + (0.5 \times 0.0126)] = 0.22$$

The conclusion is that, given the set of inventory data for the five attributes for the site to be evaluated, the probability that it is occupied turkey habitat is 22%.

Pattern recognition is a quantitative model used for predicting whether a set of habitat factors will be used by an animal. Other quantitative models can be linear or take the form of curvilinear regressions, such as the uninhibited and inhibited growth models. The HCM for the Abert squirrel in ponderosa pine is curvilinear and associates squirrel density with habitat quality and what that quality, on the average, is capable of producing (Figure 4.1). Quantitative models, while important and helpful, are difficult to develop because they require several years of data for replication of results.

4.5.4 HABITAT INDICES

The structure and life form of vegetation are the most important factors in creating wildlife food sources and cover; therefore, describing forest wildlife habitat involves measuring the amounts and analyzing the relationships between various kinds of trees, shrubs, and herbaceous vegetation. In most instances, forest wildlife biologists must use data for habitat analysis that is collected by foresters who are oriented to forest management practices and do not have a wildlife background. Traditional methods for inventorying timber stands use absolute measures of stand density, such as basal area. All the trees on a plot are measured, the areas of the individual trees are summed, and the total is stated on a unit area basis (hectare or acre). The resulting value is a measure of tree density.

Basal area (BA) is used in forestry because it is highly correlated with the growth and volume of forest stands. However, BA does not adequately describe habitat because the same BA for two different stands can result from different stand characteristics. For example, a BA of 100 ft^2/ac could be made up of 181 trees/ac of 10-in dbh or 20 trees/ac of 30-in. dbh (Table 4.4). Obviously, these two stands have quite a different cover value for a given wildlife species. In pole-sized stands, a BA of 60 ft^2/ac may provide sufficient cover for elk, whereas in a sawtimber-size stand, 90 ft^2/ac of BA are needed to provide cover (McTague and Patton 1989).

Another common forest measurement is *canopy closure*. Tree canopy closure is preferred over basal area by wildlife biologists in part because understory herbage production is directly related to the amount of light reaching the forest floor. In addition, tree canopy provides a measure of thermal cover for wildlife species such as deer and elk. However, canopy closure is generally of qualified utility because its use is largely restricted to mature stands of overstory trees.

Tree and shrub size and density are the two best measures of a stand for quantifying habitat for a particular wildlife species. However, other measures of relative size and density, such as the tree *stand density index* (SDI) developed by Reineke (1933), may prove beneficial. The SDI was seldom used until the mid-1970s, when its utility as a measure of tree density became more evident (Drew and Flewelling

TABLE 4.4
Basal Area per Acre for a Given Number of Trees per Acre

Dbh (in.)	BA (ft^2)	Basal Area/ac					
		75	100	125	150	175	200
		No. Trees/ac					
10	0.55	136	181	227	272	318	363
15	1.23	61	81	101	122	142	162
20	2.18	34	46	57	69	80	92
25	3.41	22	29	37	44	51	59
30	4.91	15	20	25	31	36	41

Note: See Section 4.10.9 (Tree Measurements).

1977). It provides a relationship between basal area, trees/acre, average stand diameter, and stocking of a forested stand. It is a quantitative measure of forest stand characteristics developed from the hypothesis that all tree species in fully stocked stands follow an identical rate of self-thinning. In fully stocked stands with complete crown closure, the following linear relationship exists between trees/acre and average stand diameter:

$$\log N = k - 1.605 \log D$$

where N is the trees per unit of land area, D is the quadratic mean diameter, k is a constant specific to tree species, and log is the logarithm to base 10.

SDI is only now beginning to be used to describe wildlife cover (Smith and Long 1987). It may prove useful in quantifying wildlife habitat once it has been further developed and related to the habitat quality of a particular wildlife species. The value of an SDI lies in the insight it provides wildlife managers in trying to assess two stands with different average tree diameters and densities. SDIs constructed for both stands may prove that they have the same wildlife habitat value (McTague and Patton 1989).

4.5.5 STAND STRUCTURE MODEL

Associating wildlife species with forest stand structure in a matrix is one way to present data in a clear and concise manner. Simple stand structure models are routinely organized and presented as manually constructed paper-based models, but the more comprehensive models that deal with large numbers of animals, food sources, cover conditions, and complex energy flows and are subject to frequent revision are best maintained in a computer. In Table 4.5, the y axis of the matrix consists of rows for food and cover. The x axis contains spaces for columns for each structural class. This type of arrangement also is ideal for developing a relation in a database.

The matrix is completed by entering a check mark in each cell for each species in each vegetation type. In some cases, an association may have a low level of resolution, such as "use" or "no use." If, however, sufficient information is available, a quantitative rating scale is used. The structure of such a model lends itself well to summarizing and communicating information on a management unit. It also serves when vegetation type and canopy closure percentages vary widely. Once a stand structure/wildlife species model has been developed, it can be augmented by linking it to tree growth and yield models suitable for evaluating both present and future stand conditions and habitat relationships.

4.5.6 WILDLIFE DATA MODELS

Considerable progress in data storage and retrieval has been made by converting the data contained in field reports to magnetic tape or disk. Concurrently, computer software to manipulate and retrieve stored data has been greatly improved. Data no longer need be in an unorganized manner but can be entered into well-organized databases for later retrieval in a systematic way. Information in wildlife databases

TABLE 4.5
Matrix for Wildlife Stand Structure Model

Wildlife Species:

Vegetation Type:

Single Story Stand

Use	:1	:2a	:2b	:2c	:3a	:3b	:3c	:4a	:4b	:4c	:5a	:5b	:5c
Food	:	:	:	:	:	:	:	:	:	:	:	:	:
Cover	:	:	:	:	:	:	:	:	:	:	:	:	:

Multistory Stand

Use	:1	:2a	:2b	:2c	:3a	:3b	:3c	:4a	:4b	:4c	:5a	:5b	:5c
Food	:	:	:	:	:	:	:	:	:	:	:	:	:
Cover	:	:	:	:	:	:	:	:	:	:	:	:	:

Code	Structural Stage	Size[a]
1	Grass-forb	
2	Seedling-sapling	(0–4.9 ft ht)
3	Pole	(5.0- to 11.9-in. dbh)
4	Sawtimber	(12- to 16-in. dbh)
5	Mature	(>16-in. dbh)

Canopy Closure[a]

a	10–40%
b	41–70%
c	>70%

Habitat Rating Scale (To Rate Each Cell in the Matrix)

5	Optimum	(1.0)
4		(0.8)
3	Moderate	(0.6)
2		(0.4)
1	Low	(0.2)
0	No value	(0.0)

[a] Size classes and canopy closure classes vary by tree species.

may be organized by guilds, life forms, or species habitat profiles or as models of various kinds in habitat relationship systems. Natural resource managers build databases because they must have ready access to information. Databases store various types of information that, when accumulated in sufficient breadth and depth, can be used to develop models and build habitat relationship systems of sufficient complexity, sophistication, and utility to greatly assist managers in better decision making.

A *database* is a collection of information organized in some rational and readily accessible fashion. In recent years, the term *database* has become almost exclusively associated with computers. But, databases are in truth a means of storing large amounts of information, whether in a computer, a library card catalog, a dictionary, or a subject file cabinet. To be useful, a database must be associated with a procedure for managing the information it contains, and this is the topic of Chapter 5, which introduces a scheme for integrating forestry and wildlife information in a data model I created. Earlier released versions of this data model were RUNWILD and later WILDHARE, which have versions still in use by national forests in Arizona and New Mexico. Data in these two relational databases have been incorporated into the FAAWN (forest attributes and wildlife needs) data model described in Chapter 5.

4.6 HABITAT EVALUATION AND MONITORING

The resource (Re) factor of Conceptual Equation 1.1 (Section 1.4.1) contains subfactors of food, cover, and water needed for a species to survive and reproduce. *Habitat evaluation* is an assessment of the quality or "health" of these factors as they affect a single species or a group of species using the same resources. Habitat and animals and their associated food chains and webs are an integral part of forest ecosystems; therefore, an evaluation of forest health, sustainability and some measure of productivity would seem to be the ideal way to evaluate habitat.

Selecting a set of standards for comparing conditions now with those that humans want to perpetuate to meet specific wildlife objectives as part of ecosystem management becomes a major problem (Figure 1.1). The problem is that past wildlife habitat management evaluation has focused on maintaining populations of huntable species or those with high aesthetic value. Only since the Endangered Species Act (U.S. Congress 1973) was passed have there been efforts to collect information on other species, such as those classed as nongame. Since then, the effort has been to obtain basic life history or distribution and abundance data and not how animals facilitate energy flow and nutrient cycling in ecosystem functioning.

4.6.1 Ecosystem Functioning

In a forest ecosystem, energy flow and nutrient cycling in food chains and food webs are likely candidates for evaluating forest wildlife habitats, but standards have yet to be determined to judge their performance. Evaluating wildlife habitats as ecosystems is a complex process and ideally should employ systems analysis techniques. This approach to evaluating an area specifically for wildlife is not well developed because of a lack of understanding of how wildlife species serve as ecosystem components. However, habitat evaluation through ecosystem management should be one goal of researchers and managers, and they should routinely be adding to the knowledge and database every new insight into the ecological processes as they become aware of this information. Management activities on the ground will also have to be viewed as experiments to determine the best management practices to meet defined objectives. Data and knowledge of processes are cumulative, so every piece of new information moves us along in understanding how ecosystems function and ultimately will provide new ecosystem evaluation techniques.

4.6.2   HABITAT QUALITY AND POPULATIONS

Estimating the numbers of animals has been discussed previously. But, how do numbers relate to habitat? A basic problem in managing wildlife is a lack of ways to determine the effects of environmental change, both natural and human induced, on wildlife populations. Once effects are known and quantified, they can be used to evaluate land treatment alternatives. One way to approach the problem is through HCMs. These models are based on the premise that wildlife populations are a function of their habitat, and habitat quality in part determines population density. The need to develop techniques for determining the present or future wildlife habitat conditions is becoming more important as forest managers develop their land use plans. The need is there because vegetation is constantly changed to benefit humans, but often at the expense of wildlife species that need it for food or cover.

The ultimate tool for evaluating habitat is a quantified system relating animal populations to different combinations of food and cover for each ecosystem: ponderosa pine, oak-hickory, northern hardwoods, and so forth. HCMs are not without problems. One criticism is that by basing population density only on quality of habitat, other factors influencing survival and reproduction are not represented. These factors are predators, hazards, disease, and competition within and between species for resources, and inherited characteristics (genetics) such as behavior in defending a territory.

This shortcoming can in part be remedied by common sense. In habitat models, vegetation is assumed to be the most important environmental factor controlling populations. By deduction, populations at some level can be related to habitat quality. For example, the tassel-eared squirrel needs trees, but how many? Certainly, neither one tree nor two, three, or four in the middle of a large clearcut is adequate. Obviously, such conditions are poor squirrel habitat. On the other hand, 10 ac with 200 trees/ac at an average diameter of 11 in. might be enough to maintain a viable population and so is something better than poor habitat. There is no doubt that all the factors of Conceptual Equation 1.1, not just resources (food and cover), have an effect on wild animal populations, but measuring and quantifying the effects has not been a practical approach for managers.

4.6.3   HABITAT SUITABILITY INDEX

A habitat evaluation procedure (HEP) to evaluate the requirements of selected species has been developed for use by the Fish and Wildlife Service (U.S. Department of the Interior [USDI] 1980). Once the species has been selected, then a habitat suitability index (HSI), based on research data and expert opinion, is calculated to compare differences between two or more areas at the same time or the relative value of the same area now and in the future. The assumption is that a numeric HSI value summarizing habitat suitability based on habitat quality with a decimal value of 0.0 to 1.0 can be developed for the selected species. One feature of the HSI is a series of equations that provide the quantitative value as

$$HSI = [V_1 + V_2 + V_3] \div 3 \qquad \text{Compensatory model}$$
$$HSI = [2V_1 + V_2 + V_3] \div 4 \qquad \text{Weighted mean}$$
$$HSI = [V_1 \times V_2 \times V_3]^{1/3} \qquad \text{Geometric mean}$$

Each of the factors (V_1, etc.) pertains to some habitat component, such as forage, escape cover, or water, with a decimal rating. A large number of HSIs have been developed for both terrestrial and aquatic habitat and can be obtained from the U.S. Fish and Wildlife Service on request. Quantified habitat models are useful but also expensive and time consuming to develop and validate sufficiently for management purposes. For this reason, ecosystem managers and biologists may have to accept something less than this ultimate tool to complete the habitat evaluation.

4.6.4 Condition and Trend

The terms *condition* and *trend* are well established in the range literature and have become part of the wildlife management terminology as well, primarily for game animals and particularly for the food element in resources. *Condition* refers to present forage composition and production in relation to the kind and amount a particular site or habitat can produce under either accepted management practices or climax conditions. The objective of a habitat condition survey is to determine whether the trend in plant development is up, down, or stable by predetermined criteria. Because of the complexity of plant communities, no single factor can be used for determining habitat condition. A host of interacting factors affect plant growth. Further, these factors never remain constant. These characteristics make it difficult to classify entire plant communities consisting of numerous species responding in different ways through time to combinations of factors.

Before an assessment of condition and trend can be undertaken, a scale factor (including time) specifying the detail and levels of resolution must be determined. Any number of habitat evaluation techniques can generally be used to make an assessment against set standards. Several methods of estimating wildlife habitat condition and trend based on key browse plants in selected areas have been developed (Dasmann 1951; Cole 1958; Patton and Hall 1966). These methods evaluate use of only key browse species (preferred plants) and do not consider the contribution of either herbaceous vegetation or other factors, such as litter or soil stability. The assumption is that the condition of the key browse species is an indication of the state of all factors contributing to habitat condition and trend. It is possible, however, that the trend in habitat condition could be downward before the shift is reflected in key browse species alone. Factors determining habitat condition and trend are

- Plant species composition
- Soil properties
- Herbage and forage production
- Degree of plant use
- Available moisture

Plant species composition is important in that there is considerable variation in forage value among and preference for different species. Also, nutritional requirements of big game animals are better met by a mixture of species. Therefore, the better habitats are those providing a variety of preferred forage species.

Reasons for including an evaluation of soils, particularly erosion, are obvious: It is the medium from which terrestrial plants obtain nutrients. Herbage production is dependent on plant density, vigor, and moisture. *Herbage* is all species of browse, forbs, and grasses. All vegetation present must be considered, even though certain plant species are not important constituents of the diet of the targeted wildlife species. In contrast to herbage, *forage* consists only of the species eaten by a given animal.

Methods of evaluating habitat condition based on ecological principles consider plant composition and use as the major criterion for judgment. Thus, a plant inventory list for the area being evaluated is a basic necessity. An essential feature of the inventory is the classification of species grouped by their successional position—whether they are *increasers, decreasers,* or *invaders* (Table 4.6). These terms refer to the ways plants react to grazing pressure by ungulates and are species specific. The terminology was developed in range management but has been extended to include wildlife species of deer, elk, and others.

Increasers are those plants that increase in numbers with increasing grazing pressure because other more desirable plants are consumed first. The opposite is true for *decreasers*. *Invaders* are plants that establish as a result of increased grazing but were not part of the original climax vegetation (native). Each condition class is related to successional climax, but it should be recognized that climax vegetation is not always the most desirable food or cover on a given site for a given wildlife species. For this reason, certain preferred or desirable plant species may indicate equal or better wildlife habitat condition than climax species.

One technique used for evaluating condition and trend for browsing animals is to estimate the degree of use of plant annual growth. Various types of transects or plots can be established, and the number of twigs browsed compared to the number not browsed provides a utilization factor (percentage) that is compared from year to year (Shafer 1963; Stickney 1966). Another approach is to enclose browse plants in small, protected plots to compare annual growth on similar plots not protected from grazing animals. This same procedure can be used for herbaceous species.

In addition to plant composition and abundance, there are other ways to determine habitat condition; the use of indicators is one way. Indicators are only as useful as the accuracy with which they reflect a cause-effect relationship. The terms

TABLE 4.6

Some Common Terms Used to Indicate Plant Value and Condition of Vegetation

Condition A	Condition B	Condition C
Decreasers	Increasers	Invaders
Preferred	Staple	Low value
Desirable	Intermediate	Least desirable
Choice	Fair	Unimportant

increaser, decreaser, or *invader,* when applied to a specific plant, are indicators. Another indicator often used is animal condition. Since animals are a product of their environment, animal condition should also be an indicator of habitat condition (cause-effect situation). Methods used to assess animal condition include indicators of productivity, such as offspring (demographic vigor), weight, and blood chemistry. One of the problems in using animal condition is that by the time the condition of the animal begins to decline, the decline in the habitat is already well advanced.

Habitat is often evaluated using a *scorecard,* which is another form of a habitat model. Scorecards help to determine present condition and relate it to statements of habitat quality. For example, a scorecard used for gray squirrel habitat in Alabama is based on five criteria in three categories with different point values (Table 4.7).

TABLE 4.7

Scorecard Used to Evaluate Gray Squirrel Habitat in Alabama

Criteria

Category I (1 point each)

  __ <50% hardwood type

  x_ 1 den tree per acre

  __ Stand age 10–40 yr

  __ Overstory 25–50% closed

  x_ Midstory 1–10%, nut- and fruit-bearing trees

Category II (2 points each)

  __ 51–80% hardwood type

  __ 2 den trees per acre

  x_ Stand age 41–60 yr

  x_ Overstory is 51–75% closed

  __ Midstory is 11–25%, nut- and fruit-bearing trees

Category III (3 points each)

  x_ 81–100% hardwood type

  __ 3 den trees per acre

  __ Stand age > 61 yr

  __ Overstory > 76% closed

  __ Midstory > 25%, nut- and fruit-bearing trees

Total Point Value

  1–5 is poor

  6–10 is fair

  11–15 is good

Source: U.S. Department of Agriculture (USDA), *Wildlife habitat evaluation program (WHEP).* Forest Service, Southeastern Area, State and Private Forestry, Region 8, Atlanta, GA, 1974.

From this example, the following information is available: Hardwood composition is 90%; the area contains one den tree per acre in stands that are about 55 yr old, with an overstory 75% closed, and a 5% midstory of nut- and fruit-bearing trees. The rating is determined from total point value, where 1–5 is poor, 6–10 is fair, and 11–15 is good. The total point value of the three categories for this example is 9, so the rating is fair.

Another evaluation technique frequently used in the West is the classification of shrubby plants (browse) into form and age classes (Table 4.8). Availability takes into account the height of the shrub in relation to how high an animal can reach to obtain leaves and stems. The degree of hedging is a subjective judgment based on experience. *Hedging* is an estimate of the effect of browsing on the regrowth of stems after they have been severed. Several years of continued use cause the stems to clump.

The classification of browse plants into age classes varies according to the species. Ages are often assigned according to main stem diameter size; for example, young might be in the 1/8- to 1/4-in dbh class. Plants are considered decadent if more than 25% of their crown is dead. Plants classified by form and age class along a permanent transect are compared with previous measurements, usually at 3-yr intervals, to arrive at an evaluation for deer, elk, or cattle use of an area. The examples given are just two of the many types of scorecards (habitat model) that have been and are still being used by wildlife biologists to evaluate habitat. While the criteria for each scorecard vary, they almost always are based on a system of total points to arrive at a rating using adjective statements (poor to good, etc.).

TABLE 4.8

Classification of Browse Plants by Form and Age Class

Form Classes

1. All available, little or no hedging
2. All available, moderately hedged
3. All available, severely hedged
4. Partly available, little or no hedging
5. Partly available, moderately hedged
6. Partly available, severely hedged
7. Unavailable
8. Dead

Age Classes

S = Seedling
Y = Young
M = Mature
D = Decadent

Source: Patton, D. R., and J. M. Hall. *J. Wildl. Manage.* 30:476–480, 1966.

4.6.5   MONITORING HABITAT CHANGE

Monitoring is the process of checking, observing, or measuring the outcome of a process or the changes deviating from some specified quantity, such as in the scorecard example. Detecting change has always played a significant role in wildlife management, whether in a low-resolution population analysis or some higher level of measurement aimed at determining habitat condition or trend. Monitoring has taken on a new dimension since it has been incorporated into resource legislation and regulation. As a result, there has been increased attention given to new approaches to monitoring.

The techniques for monitoring both habitat and populations, or indicators of the two, have essentially already been developed, so at this writing no new avenues are evident other than improving the accuracy and efficiency of current techniques, including remote sensing, geographic information systems (GIS), and database management systems (DBMS). Managers needing to monitor an area or the outcome of a management plan are confronted with the same old problems of how much detail is needed to detect change and the cost of obtaining this level of detail. Monitoring is a feedback mechanism involving five steps (Salwasser et al. 1983):

- Set objectives
- Identify actions and impacts to be evaluated
- Collect and analyze data
- Evaluate results
- Make corrections in the system

The ultimate goal should be to set in motion a detection network to monitor the resource (Re) factor of Conceptual Equation 1.1 indirectly through monitoring habitat condition and trend. While the techniques to achieve this have already been developed, although not always refined, the means of constructing an integrated system to monitor all forms of wildlife has not. The effort to correct this shortcoming has led to some interesting and fruitful approaches in systems analysis.

The most direct way to detect changes affecting wildlife is to monitor vegetation quantity (hectares, acres) and quality (succession) over time. This can be done for each management unit on a map documenting acreage and location of vegetation by successional stage. Area size, stand size, and location are the factors monitored over time. Location is as important to monitor as is area size for it depicts the juxtaposition of food and cover.

Verner (1983) has developed a flowchart depicting the relationships among elements of an integrated system for monitoring wildlife populations (Figure 4.8). While the boxes cannot identify the predictive tools used in action items to be implemented and goals to be evaluated, getting the subsystems within each of the boxes to work will be a long-term process. The flowchart, nonetheless, is a realistic approach to solving the problem of an integrated monitoring system.

4.7   INTEGRATED MANAGEMENT

Integrated resource management planning (IRMP) is the process that puts ecosystem management to work at the project level. The importance of the IRMP process

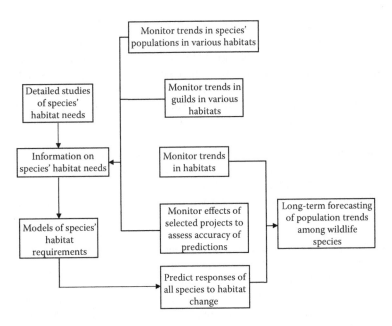

FIGURE 4.8 A systems approach to monitoring multiple wildlife populations. (From Verner, J. *Trans. N. Am. Wildl. Nat. Resources Conf.* 48:355–366, 1983.)

is in the collective thoughts of all the interdisciplinary team (ID team) members to develop management alternatives that are based on the best information available at the time. However, for the IRMP process to work, as one part of maintaining sustainable forest ecosystems, the ID team should include members outside their agency.

What does the general public want from state and federal forestlands? In the final analysis, it all depends on each individual's education, experience, and view of the world, country, and state. If the situation of a healthy, sustainable forest over a landscape can be attained, then probably all of the amenity and commodity desires of people will be met. The natural components in the landscape are linked one with the other in areas sufficiently large to serve the public. The human component is linked to the ability of the system to produce and the willingness of people to accept that limits exist and that some components, such as wildlife habitat and populations, old forest conditions with large trees, and available timber volume, will change in time and space.

4.7.1 ADAPTIVE MANAGEMENT

In this chapter, the idea has been expressed, in relation to ecosystem management, that vagueness and uncertainty are the normal result when trying to predict the future. Sometimes when there is uncertainty, a risk analysis is appropriate and will help in making a decision (Marcot 1986). Because of uncertainty about how a particular management decision might affect future outcomes, managers need some process that will allow for adjustments over time.

In the 1970s and 1980s, C. S. Holling (1978) and Carl Walters (1986) began to develop a process that has now come to be known as *adaptive resource management*. The outcome of these publications led to the thesis that scientific understanding will come from the experience of management as an ongoing, adaptive, and experimental process, rather than through basic research or the development of ecological theory. Walters explained that there are three essential ingredients in adaptive resource management:

- Modeling to determine uncertainties and generate alternatives
- Statistical analysis to determine how uncertainties are likely to propagate over time in relation to policy choices
- Formal optimization combined with game playing to seek better problem choices

The ideas of adaptive resource management have been in place for many years; for example, in the 1940s and 1950s state resource agencies were struggling with setting hunting and fishing seasons and bag limits. The process involved using the limited biological knowledge that was available at that time along with the experience of "game wardens" and "conservation officers" to arrive at a set of regulations. These regulations were then reviewed by the public, sometimes in hotly debated sessions, with the result that often public pressure dictated the seasons for the next year.

Once regulations were established, then the affected agency had a year to observe the outcomes of the various seasons and limits to repeat the process for the next year. Ultimately, the process resulted in an acceptable set of regulations with only minor changes from year to year. Even today, the process is about the same but with more concerns added, such as those for endangered and threatened species, old forest conditions, forest health and sustainability, riparian vegetation, timber harvesting, and livestock grazing. Adaptive resource management will gain more acceptance and be used regularly in the future as resource agencies move from the local project and single species to local landscapes and multispecies approaches to management.

4.8　PLANNING

The land base for resource management is diminishing due to the increasing number of cultural landscape features needed by an increasing human population (remember what Pogo said; see page xv). Present conflicts in allocating land to wildlife habitat can only intensify with human population growth. The problems of allocating resources are compounded by the fact that different wildlife species have different food and cover requirements; this requires setting aside a variety of habitats to accommodate these species. The allocation of habitat to forest wildlife takes place in the planning process at the forest level and is based on the mandated principles of multiple use. A further complication in the allocation process results from legislation requiring the maintenance of viable populations of all native vertebrate species in the planning area. This legislative mandate is a consequence of the biological fact that virtually all animal species are included in and vital to the functioning of an ecosystem.

The balance among resource use, timber, water, livestock, wildlife and ecosystem functioning, energy flow, and nutrient cycling reflects the system of political and biological decisions that result in forest plans. The political decision determines how much acreage is designated for a given habitat and how many animal species will be maintained in the same habitat. The biological decision determines the effect of the political decision on animal and plant species in the management unit and provides alternatives whenever conflicts arise.

Land management planning clearly is not in the *centrum* of any animal species since this factor does not directly affect the chance of an animal to survive and reproduce. But, land management planning does affect the six factors of Conceptual Equation 1.1, so it properly belongs in the *web* from a biological or ecological viewpoint. Yet, if land management planning has been made by virtue of legislation, a factor of such importance that it requires considerable time by managers, then it takes on the characteristics of a seventh factor in the centrum regardless of the dictates of logic.

4.8.1   Laws and Policy

Land use planning depends in the first instance on legislative authority empowering a federal or state agency to act and in the second instance on a policy formulated by that agency to use its authority. A long list of legislative statutes provides the authority for various U.S. agencies, the Fish and Wildlife Service, the Bureau of Land Management, and the Forest Service to manage natural resources, including fish and wildlife on public lands, and to cooperate with state agencies in conducting these management activities.

While federal agencies have been authorized in a few cases to manage populations (migratory species) on public lands, the generally accepted division of responsibility is that habitat management on federal land is the responsibility of federal agencies, but population regulation is controlled and enforced by state agencies under state law. Because this book is particularly concerned with forest wildlife management across the continent, it is appropriate to summarize a few of the major laws controlling federal land management agencies.

4.8.2   Fish and Wildlife Coordination Act

The Fish and Wildlife Coordination Act (U.S. Congress 1934) provides that fish and wildlife resources receive equal consideration with other resources in water development projects. It authorizes and directs the U.S. Fish and Wildlife Service and the relevant state agencies to coordinate water development programs with the other land management agencies and to encourage the installation of fish and wildlife measures to protect or enhance habitats.

4.8.3 MULTIPLE-USE, SUSTAINED-YIELD ACT

The Multiple-Use, Sustained-Yield Act (U.S. Congress 1960) recognizes and clarifies the U.S. Forest Service authority and responsibility to manage fish and wildlife. It defines multiple use as

> the management of all the various renewable surface resources of the National Forest so that they are utilized in the combination that will best meet the needs of the American people; making the most judicious use of the land for some or all of these resources or related services over areas large enough to provide sufficient latitude for periodic adjustments in use to conform to changing needs and conditions; that some land will be used for less than all of the resources; and harmonious and coordinated management of the various resources, each with the other, without impairment of productivity of the land, with consideration being given to the relative values of the various resources, and not necessarily the combination of uses that will give the greatest dollar return or the greatest unit output.

4.8.4 SIKES ACT

The Sikes Act (U.S. Congress 1967) directs the secretary of agriculture and the secretary of the interior, in cooperation with state agencies, to plan, develop, maintain, and coordinate programs for the conservation and rehabilitation of wildlife, fish, and game. It provides for fish and wildlife conservation programs on federal lands, including authority for cooperative state-federal plans and authority to enter into agreements with states to collect fees for funding programs identified in those plans.

4.8.5 NATIONAL ENVIRONMENTAL POLICY ACT

The National Environmental Policy Act (NEPA) (U.S. Congress 1969) requires that a systematic, interdisciplinary approach be employed in planning actions that may have an impact on the human environment. It mandates the preparation of a detailed statement on the effects of legislative proposals or other major federal actions significantly affecting the quality of human environments. Wildlife, fish, and their habitats must be fully considered in project planning and decision making.

4.8.6 ENDANGERED SPECIES ACT

The Endangered Species Act (U.S. Congress 1973) provides the means by which the ecosystems on which endangered and threatened species depend may be conserved. It provides a program for the conservation of such endangered and threatened species and authorizes such steps as may be appropriate to achieve the purposes of the international treaties and conventions. It establishes federal policy enjoining all federal departments and agencies to conserve endangered and threatened species.

4.8.7   FOREST AND RANGELAND RENEWABLE RESOURCES PLANNING ACT (RPA)

The Forest and Rangeland Renewable Resources Planning Act (U.S. Congress 1974), directs the secretary of agriculture and the U.S. Forest Service to make an inventory and assessment of the present and potential renewable resources of the United States every 10 years. Wildlife and fish, as part of the renewable resources of the United States, are included in the inventory and assessment, as is an analysis of present and anticipated uses, and the demand for and supply of renewable resources is to be included.

4.8.8   NATIONAL FOREST MANAGEMENT ACT (NFMA)

The NFMA (U.S. Congress 1976) provides for balanced consideration of all resources in U.S. national forest land management planning. It directs the secretary of agriculture to protect and improve the future productivity of all renewable resources on forestland, including improvements of maintenance and construction, reforestation, and wildlife habitat management on timber sales. The act specifies that habitat diversity includes variety, distribution, and relative abundance of plant and animal communities in a multiple-use context.

The RPA and the NFMA have had a particularly profound effect on the management orientation and activities of the U.S. Forest Service. Both acts required major changes in the management direction of the agency and demanded a massive commitment of human and financial resources to complete the assessment on a national scale and then prepare accompanying management plans at a forest level in a timely way.

The RPA and NFMA called for cooperation with all the federal land management agencies as well as those of the 50 states and Puerto Rico. The mandated changes have, in turn, resulted in marked changes in the way state natural resource agencies manage their activities. Such a complete assessment of natural resources never before had been undertaken. What was to be assessed, how was it to be done, and how it was to be presented required extraordinary consultation and input from a wide variety of sources: federal, state, and private. The end product was *An Assessment of the Forest and Rangeland Situation in the United States* (U.S. Department of Agriculture [USDA] 1981). Additional assessments are to be completed at the beginning of each decade.

4.8.9   GOVERNING POLICIES

Legislative authority to manage does not automatically result in a management strategy by a state or federal entity. The agency must develop a policy to guide management in the allocation of staff, finances, and other resources to achieve their goals. *Policy*, in short, is the setting of a course of direction. The first step in formulating policy is the preparation of a statement that summarizes the objectives to be realized and the process by which they will be achieved. Policy statements must be clear and easily communicated through a chain of command. An example of a policy statement is found in a U.S. Department of Agriculture memorandum (1982), which begins as follows:

> The purpose of this memorandum is to state the policies of the U.S. Department of Agriculture with respect to management of fish and wildlife and their habitats and to prescribe specific actions to implement the policies.
>
> ... It is the policy of the Department to assure that the values of fish and wildlife are recognized, and that their habitats, both terrestrial and aquatic, including wetlands, are recognized, and enhanced, where possible as the Department carries out its overall missions.
>
> ... A goal of the Department is to improve, where needed, fish and wildlife habitats, and to ensure the presence of diverse, native and desired non-native populations of wildlife, fish and plant species, while fully considering other Department missions, resources, and services.
>
> ... On public lands administered by the Department, habitats for all existing native and desired non-native plants, fish, and wildlife species will be managed to maintain at least viable populations of such species.
>
> ... This will be accomplished through the forest planning process in response to targets identified in the Forest and Rangeland Renewable Resources Planning Act program and public issues and concerns brought up in the planning process, consistent with available resources. Habitat goals will be coordinated with State Comprehensive Plans developed cooperatively under Sikes Act authority and carried out in forest management plans with State cooperators. Monitoring activities will be conducted to determine results in meeting population and habitat goals.

This policy statement continues with sections dealing with threatened or endangered species, economic losses from plant and animal pests, authorities to implement the policy, and responsibilities for implementation and coordination.

The secretary of agriculture is a cabinet position responsible for several agencies. Policies for each agency must be interpreted from the secretary's memorandum. The subsequent interpretation of the secretary's policy at a lower level is found in the Forest Service statement (USDA 1984) on fish and wildlife, which contains the following:

> Serve the American people by maintaining diverse and reproductive wildlife, fish, and sensitive plant habitats as an integral part of managing National Forest ecosystems. This includes recovery of threatened or endangered species, maintenance of viable populations of all vertebrates and plants, and production of featured species commensurate with public demand, multiple-use objectives and resource allocation determined through land management planning.

The U.S. Forest Service policy statement interprets and adds to the policy of the department due to the fact that the Forest Service has authorities that other departmental agencies do not. The intent in policy statements is to clarify and strengthen objectives and goals at each lower level of administrative responsibility. The aim of well-written, succinct policy statements is to clearly communicate direction so that each employee has simple guidelines for the performance of his or her tasks and can adequately represent the agency to the public.

Interpretation of policy can vary when political appointees change from one administration to another. As a result, what is policy today may not be policy tomorrow, depending on how the regulations are interpreted and how Congress reacts to national issues.

4.8.10 WHY PLAN?

Planning at some level is required by law for most state and federal natural resource agencies. It is necessary not only to discharge the responsibilities of the agency but also to properly respond to a world of change and competition for financial resources and space. Change can be orderly to meet specific goals and objectives, or it can be random and nondirected. Human-induced change, to be orderly, must be controlled in some fashion. The planning process establishes criteria, principles, and procedures for control and documents the sequence in which events should happen. The planning process and the resultant plan are means to an end or goal.

The planning process is the forum in which competing demands and objectives are finally integrated to collect the available resources into a practical working tool for management and staff. While the role of wild animals in the functioning of an ecosystem is not well documented or thoroughly understood, it is already clear in a general way that they are crucial in nutrient cycling and energy transfer. By virtue of the ecological role wild animals have as an ecosystem component, the wildlife management function in an agency should be placed on a funding level equal to that which other natural resource functions, such as timber, water, and livestock, have had for many years.

To adequately manage all plant and animal populations at any unit level requires a detailed plan of action. Employing the knowledge and insights of a variety of professional personnel, such as foresters, ecologists, and wildlife biologists, in the planning process is a necessity. The trend away from single species or single resource management to ecosystem management is a positive approach to ensuring that few, if any, animal species go below viable population levels as a result of human activities unless done as a planned objective. One exception to multiresource management and planning is managing for an endangered species when single species management is biologically sound and legally justifiable.

To compete for dollars and space, to cooperate in the formulation and selection of alternatives, and to integrate wildlife objectives into ecosystem management plans, forest biologists must have access to the kind of information outlined in the preceding chapters. Now, as never before, biologists as planners must investigate and understand how wildlife interacts with and affects other ecosystem components. Ecosystem managers have to ensure that all the resources are fully considered and provided for in the planning process.

4.8.11 FACTORS TO CONSIDER

The wide-ranging and extensive body of laws and regulations relating to natural resource management directly affect both the planning process and the plan itself. As in preparing inventories, plans must be programmed at different scales, depending on whether they are for national, regional, or local use. The scale of the plan will dictate not only the kind and detail of the information included but also the process and format. Plans at the national and regional levels must be presented in a standardized format to facilitate understanding by management agencies, conservation organizations, and environmental groups.

Most natural resource agency plans are one of two types: strategic or operational. *Strategic plans* guide an agency or organization in meeting long-term goals and are stated in economic, technical, and political terms. Strategic plans are often referred to as *long-term plans. Operational plans* guide specific actions and allocation of resources on a shorter timescale, such as a single year, several months, or weeks. An operational plan deals in terms of specific management units or parts, specific ecosystems and particular organisms therein, the relationships between the environmental characteristics of the unit and their organisms, and the ways in which these relationships are to be maintained or allocated. Operational plans are known in natural resource professions as management plans. While national and regional plans are important, most field biologists are involved in preparing the local plan or management unit plan, so the focus here is on this kind of planning.

4.8.12 OPERATIONAL PLANS

All wildlife management plans have certain elements in common that must be included if the plan is to be a useful working document. Two of these common elements, both of high priority in the process, are the setting of goals and objectives. *Goals* are broad statements of central themes that set management direction. They are seldom quantifiable and generally have no time limit. *Objectives* are specific statements of activities to be undertaken to realize the goals set for a management unit. Objectives can be measured or accounted for through a given time. One approach, known as management by objectives (MBO), fits well with a planning methodology widely adopted by many federal and state agencies. The technique depends on adopting specific objectives and determining how they will be measured. Agency personnel are then expected to account for the completion of each objective in terms of the measurement stipulated.

The element of time was discussed in relation to succession, forest rotation, and cutting cycles. All these are factors common to management plans and must be included, but another time factor, planning period, is an important factor and must be part of the plan. Time checks are routinely built into management plans at the end of a given period; for example, every 5 years they provide a major checkpoint to evaluate accomplishments and set new directions if required. Managers clearly need to know the starting date when a plan comes into operation, but equally important is the date when the plan is either completed or to be revised. An outdated plan is as useless as no plan. The planning period is often dictated by state or federal law or regulations, as is the ability to revise a plan when such a need arises. Revisions can result from advancing technology, to correct errors or incorporate new data, or to address the expressions of public concern about a management activity. While the life of a plan may be clearly stated, it should not be so rigid that revisions and updates cannot be made as conditions change.

4.8.13 Checklist for Background Information

Management unit plans require considerable data, much of which may already be available but which should at least include the following:

Geographic data: Geographic maps are the basic documents needed to locate boundaries and forest vegetation types.

Historical data: These would be history, farming, logging, mining activity, and so forth.

Wildlife census data: These date provide information on populations, both current and historical.

Game kill records: These are often used to supplement historical census data.

Aerial photographs: These are used to associate vegetation with landscape features when transferred to topographic maps.

Satellite multispectral images: These images are often useful in establishing land use patterns surrounding a management unit.

Range condition and trend surveys: These provide information on herbage production and utilization.

Soil surveys: These surveys are used to establish vegetation potential and productivity.

Allotment analysis reports: These reports establish past and present livestock use of an area.

Timber inventory records: Stocking and stand tables are needed to estimate current and future growth and yield.

Vegetation type maps: The maps are keyed to identify food and cover areas.

Exclosure location and data: These data help to establish plant successional trend and use patterns.

Water inventory: The location of seeps, springs, streams, and small catchments helps determine the need for water development projects for habitat improvement.

Recreation facilities: This information is used to determine human impact on the management unit. Road and trail information identifies access routes that have an impact on the management unit.

Much of the information listed will presumably have been collected by various agencies in the course of their management activities, so often this planning phase is simply a matter of bringing information together and checking for missing data. It is critical that, as these data are accumulated for inclusion in a new plan, that they be simultaneously entered into a database so that complete up-to-date information is always available for future use.

4.8.14 Decision Support Systems

In response to the needs for a national and regional system of planning, the Forest Service adopted a linear programming model called FORPLAN (Mitchell and Kent 1987). FORPLAN (forest planning) was designed to provide management alternatives by presenting various levels and combinations of goods and services that a forest can provide. While FORPLAN greatly assisted the Forest Service in meeting its legal responsibilities under the NFMA for planning at the national and regional levels, it provided little planning assistance at local levels. But, as stated, resource management planning is usually done at the forest subunit level. To assist managers at this level decision support systems (DSS) are used to provide alternatives for the planning and decision-making process on small units of land. This type of analytical tool is appropriately titled because DSS link a variety of different modules to form a systematic planning structure. The modules can be linked as individual forest models, such as matrices, plant and animal databases, GIS, or linear program optimization models.

Before FORPLAN, the planning goals and objectives of resource agencies were simple, requiring only minimal amounts, types, and complexity of data. Therefore, information generally could be analyzed and presented without the aid of complicated automated systems. This is not the case today because types and amounts of information needed to meet the requirements of the natural resource laws, and agency regulations implementing these laws, are sufficiently complex that new ways of analyzing data had to be developed and incorporated into the planning process. FORPLAN was the primary tool used in developing the forest plans required under the NFMA.

FORPLAN software is used with data supplied for each analysis area. The data must be developed for management prescriptions and estimated costs, together with the outputs, effects, and benefits for each prescription or combination of goods and services. This information is then used to generate matrices of managerial options and associated benefits, costs, and so forth. The end product is a listing of the optimum combinations of prescriptions and associated benefits. Needless to say, the final decision regarding which combination of prescriptions is selected continues to be the responsibility of skilled natural resource managers. FORPLAN is not the only management planning system available to natural resource managers; it is representative of the types of tools that are becoming available for planning, decision making, and decision support.

4.8.15 PROJECT IMPLEMENTATION

As a result of the NFMA, some Forest Service regions developed a project implementation process (PIP) to ensure that all projects and undertakings that generated human impacts, such as mining, range improvements, road construction, and timber sales, were carried out in accordance with previously developed plans for each forest (USDA 1988). The PIP process grew out of and reflects the adoption of the policy to manage the national forests on the basis of an integrated resource management (IRM) approach. IRM is based on the recognition that all natural resources are connected by an intricate web of interrelationships. As a result, any activity undertaken with one resource affects other resources in the forest simultaneously. Because the interrelationships between natural resources are extraordinarily complex and not always well understood, the effects of a management activity on various elements of the forest ecosystem are difficult to predict.

The IRM process uses an interdisciplinary team for project design to predict the effects of a management activity. The team identifies all the resources involved, defines the interrelationships, and predicts the effects and impacts of the project on all the identified resources. To facilitate the work of the team in the project design, each project must have specific, quantifiable objectives. When the plan is in place with all impacts identified, the land manager can then implement the project and use the interrelationships between the resources to accomplish objectives.

The interdisciplinary team prepares a brief report to determine whether the proposed project is technically and economically feasible. Based on the report, line officers decide whether to proceed with the project. The team must also develop a reasonable range of alternatives so that responsible officials can select those to be implemented and then determine what NEPA documentation is appropriate. Finally, the project is completed and evaluated for the success or failure of project design.

IRM is another powerful tool for planning and implementing the objectives of a plan that can have an impact on a variety of natural resources. It offers the wildlife biologist, as part of an interdisciplinary team, an unsurpassed opportunity to reduce the negative impacts on or to improve the habitat of a specific species or forest wildlife in general. IRM and its associated PIP coupled with a DSS ensure natural resource managers and the public that the most modern and scientific methods are being used for planning and decision making.

4.9 INVENTORY

The number of techniques available to inventory wildlife species and habitat is sufficiently large so that managers can select from a variety of tools. While the availability of a large number of techniques may be desirable, it can create a problem in selecting one that is cost effective and biologically sound. As a result, biologists must spend considerable time planning the job before starting fieldwork for an inventory.

Since passage of resource management legislation in most of the developed countries, a new emphasis on inventory has resulted because several different types of data are needed to fulfill all the decision-making requirements. The emphasis is now on inventorying all resources at the same time, in other words, a multiresource inventory. Inventories are basic to management and so must be planned carefully. Scientific management is impossible without an inventory, but too often an inventory is made before adequate planning and objective setting have been undertaken.

Planning must start by setting objectives. Experience has shown that without clear, well-defined objectives, one of two things happens: either too much or too little data are collected or the wrong types of data are collected. Either results in the loss of both time and money. The technique must fit the resources available to do the job.

4.9.1 MULTIRESOURCE INVENTORY

Once a decision has been made to conduct an inventory, then one task needing early attention is to determine if data can be collected as a part of another resource inventory, such as for timber or watershed management (Lund et al. 1978). If no such inventories are planned, it will have to be initiated as a single-purpose inventory, such as for wildlife. In short, do not collect data if they are already available. Make use of existing sources such as timber surveys, allotment analysis, recreation plans, and watershed and soil surveys, even if the data need interpreting or revision. In any case, the actual data collection must be consistent. Use of the following guidelines will help to meet this goal:

- Collect data using the same methods
- Collect data at the same time of year
- Collect data on the same plots if the sample was systematic as in previous inventories

Use a field form that forces data entry into an allocated space for a given topic in the order it will be entered into a computer. This procedure reduces the possibility of mistakes by a data entry operator. Another alternative is to use an electronic field recorder programmed for data transfer to a microcomputer and into a DBMS. *Field data recorders* (FDR) are portable, battery-operated microcomputers. Most of the new FDR can be programmed with rules allowing only data with certain characteristics to be entered into storage. The advantage of some types of recorders is that the data can be transmitted electronically by telephone at night to an office computer.

4.9.2 CATEGORIES OF INVENTORY

Category of inventory has to do with purpose and how the information collected will be used. In general, there are three hierarchical categories of inventory: national, regional, and management unit. These categories are not all inclusive, and there can be overlap between them; they are not the only categories that can be identified. However, they are intuitive because of the commonality with geographical and political boundaries.

National inventories usually provide limited information consisting of lists of resources used for formulating legislation, determining policy, setting management and research priorities, and supporting budget presentations. Information collected just for a national inventory generally cannot be disaggregated to a lower category. Regional inventories typically focus on a geographical area or major vegetation types that cut across political boundaries. This type of inventory is used to develop programs, allocate funds, ensure distribution of resources, and provide for regional diversity. The information becomes part of a planning document that can be revised and updated to conform to changing conditions.

The third category of inventory is a management unit that will vary; it can be as large as a national forest or as small as a watershed. Management unit inventories are the basis for management decisions and most often include detailed resource information. Information collected at the management unit level generally contains too much detail for all of it to be aggregated to higher categories. However, in almost all cases some information is suitable for regional use.

The categories of inventory are additive from the local level to a national level, such as national, regional, and local. This scheme has many advantages, particularly if all the data are collected at the same detail, format, and consistency at each level. A framework for an inventory that considers multiresources would include

- Tree size: diameter and height
- Tree density: distribution, layers
- Stand size
- Natural openings
- Created openings
- Plant species: abundance, frequencies
- Understory vegetation: plant frequency, abundance
- Forage production: understory, open area
- Disturbance: roads, trails, campgrounds
- Water sources
- Topography
- Down material: logs, stumps, debris
- Snags: density, size, condition
- Evidence of wildlife species: documentation

4.9.3   INTENSITY OF INVENTORY

The intensity of an inventory must be controlled by the purpose and objective of the inventory, but too often it is controlled by the money available. Ideally, the level of detail increases with the level of intensity, from low to high. Low-intensity inventories consist largely of lists of species, grouped by some characteristic of interest such as threatened and endangered, game and nongame, taxonomic group, or economic value. Low-intensity inventory information is frequently used at the national level, but it is also the first step in a high-intensity inventory for a management unit. Data for low-intensity inventories are largely derived from literature surveys or expert opinion.

A moderate-intensity inventory builds on the species list and classification by adding information on where a species feeds and breeds. It also can include an estimate of population abundance using general terms of low, moderate, or high and a population trend of increasing, decreasing, or stable. Data for moderate intensity inventories are derived from literature reviews, expert opinion, and limited field surveys.

High-intensity inventories are the most detailed and most expensive and take the greatest amount of time to complete. Data are collected by field crews who have knowledge of the animals and their habitat. Data include plant species composition and abundance, animal population characteristics (density, sex and age ratios, etc.), and quantity and quality of habitat. If data are collected properly, they can be aggregated with other management units to provide information for regional and national purposes. The intent and scope of high-intensity inventories must be carefully planned, priorities defined, precise cost estimates established, and the right types of data clearly identified.

An inventory of resources is the first step in the management process, and without an inventory there can be no management except by default. The manager, when considering category and intensity of inventory, can select from several combinations, but some are not as practical as others because the detail is not necessary or the cost is too high.

4.9.4 MAP SCALE

After a selection has been made for category and intensity of inventory, the next major job is to select a working map at a scale that provides an acceptable level of resolution. *Scale* is the ratio of actual distance on the ground to the same distance on the map. Scale is a representative fraction by which the map and ground distances are shown in the same units, such as 1:63,360, which means that 1 inch on the map equals 63,360 in. (1 mi) of the same units on the ground.

A map with a scale of 1:24,000 is the standard for a 7.5-minute latitude (angular distance north or south from the equator of the Earth measured through 90 degrees) and 7.5-minute longitude (great circles running through the North and South Poles). Each quadrangle (square sheet) covers approximately 40,215 acres. Large-scale maps, such as 1:7,920 or 1:24,000, are useful when small areas need to be delineated with detailed information. Intermediate-scale maps range from 1:50,000 to 1:100,000, cover larger areas, and are suitable for land management planning (Steger 1986). Maps of 1:250,000 to 1:1,000,000 are small scale and cover very large areas on a single sheet.

Scales of 1:7,920 and 1:15,840 are often used because they provide acceptable levels of resolution for different levels of inventory, but they are not the only scales used by modern map makers (Table 4.9). These two scales are also commonly used in aerial photography in the United States because conversion from inches, miles, and acres is readily made, and detail from photographs is easily transferrable to the maps. The outline of the management unit drawn on the map provides the basic reference for sampling and determining area.

In the United States, the National Cartographic Information Center can provide information about maps, charts, aerial and space photographs, satellite images,

TABLE 4.9
Common Map Scales

Scale Ratio	Ft/in.	In./mi	Ac/in.^{2}
1:500	41.667	126.72	.0399
1:1,000	83.333	63.36	.1594
1:2,000	66.667	31.68	.6377
1:5,000	416.667	12.67	3.9856
1:7,920	660.000	8.00	10.0000
1:15,840	1,320.000	4.00	40.0000
1:20,000	1,666.667	3.168	63.7690
1:24,000	2,000.000	2.640	98.8270
1:31,680	2,640.000	2.000	160.0000
1:63,360	5,280.000	1.000	640.0000

geodetic control, and digital and other cartographic data. A catalog is available from the National Cartographic Information Center, U.S. Geological Survey (507 National Center, Reston, VA 20192-0003).

4.9.5 GEOGRAPHIC INFORMATION SYSTEMS

A GIS is a computerized means of making base maps with overlays. Every forest biologist appreciates the value of overlays in decision making because of the many hours spent in using colored pencils to code habitat characteristics, delineate hunt unit boundaries, and locate water developments on acetate film. Overlays present a simultaneous visual interpretation of the several factors involved that would otherwise be difficult to explain in written or oral form.

GIS maps built from data transferred to a computer electronically, by tracing on the map with a handheld digitizer to create overlays, require but a fraction of the time expended using the manual method. A digitizer converts line data to points that can be connected and later converted back to lines on a map. One major advantage of a GIS is that revisions can be made without revising the entire map. Another advantage is that a combination of overlays can be used to show where characteristics coincide.

When a GIS is coupled with a DBMS containing detailed information and relates it to items on the overlays, the total system becomes a powerful management tool. For example, a base map of boundaries of a group of timber stands is prepared by GIS. In an associated DBMS, the characteristics of the stands are recorded (species, acres, tree density, quadratic mean diameter, etc.). The DBMS also contains another file with quantified data on habitat conditions for a given wildlife species. Knowing that the stands and habitat characteristics have been defined, overlays can then be plotted that identify stands with the defined habitat characteristics. If the DBMS also contains a file describing tree growth and yield, then stands can be manipulated in the computer to model habitat conditions for future dates and alternative management practices.

4.10 SAMPLING

After the map scale has been selected, a sampling design can be superimposed over the management unit delineated to show location of plots or points of a transect. Sampling design is a detailed and complex topic, and it cannot be adequately presented in a few paragraphs. There are, however, several useful sampling and plot guidelines that should be followed:

- Use nested plots for sampling different resources in the same area
- Use circular plots to reduce border area
- Use a systematic sample with a random component to reduce the cost of sampling
- Make a preliminary sample to determine variance to estimate the number of plots needed for a given confidence level

4.10.1 Types of Sampling

In general, there are three basic sampling systems that can be used in vegetation inventory: point, plot, and plotless. *Points* are commonly used for sampling herbaceous plants and grasses at given distances along a permanently established line (transect). Only the species under the point is recorded. Some point methods use a small ring (in square inches or centimeters) and count the number of plants within the ring. This type of sampling has been used primarily in grassland vegetation.

Plotless sampling at a point is used primarily for trees. The sample is obtained by using a calibrated prism to decide whether a tree is to be counted. This type of sampling is fast and accurate if the sample is sufficiently large to overcome variance. The point can be located on an established transect line or selected at random.

Plots are the most common system in use for all types of sampling of vegetation or animals. Plots have dimension, so the density of organisms is easily calculated. Vegetation sampling to measure the attributes of different life forms of trees, shrubs, and herbs can be done using a single plot or nested plots. *Nesting* means that a large plot may enclose a smaller plot (Figure 4.9). Nested plots are advantageous because they incorporate different plot sizes at one location for sampling different factors. The center of the large plot may or may not serve as the center for smaller plots. Smaller plots can be systematically located within the larger plot. An alternative to nesting is to cluster the subplots around the outside of the primary (largest) plot.

A *transect* is a graduated line for use in sampling with points or plots. It can be a line with a width of centimeters or meters or a line without dimension with plots of any size or shape at given intervals. Many combinations of sample systems are possible, and because of this, sufficient time must be taken to determine the correct system for the job to be done.

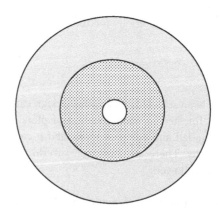

Large = 0.1 ac (trees, logs, stumps, etc.)

Medium = .01 ac (shrubs, small trees, wildlife sign)

Small = 9.6 ft^2 (herbaceous vegetation)

FIGURE 4.9 Nested plots are useful in habitat inventory.

4.10.2 SHAPE AND SIZE OF SAMPLE UNITS

When sampling vegetation in habitat surveys, data are often collected from within circular plots on a transect. Circular plots have proved satisfactory in sampling because they are the easiest to use when deciding what is to be counted if an organism is partially in a plot. This is because the length of the circumference of a circle is smaller than the length of the sides of a square of the same size, so fewer border decisions have to be made.

The *plot size* selected for sampling vegetation must be based on the spatial characteristics of the vegetation. In general, the radius for the sample plot is equal to the average distance between the items to be measured. One way to determine plot size is to conduct a nested plot study and then construct a density area curve (a modification of the species area curve) (Mosby 1963). For trees, four or five different plot sizes are nested, and trees are counted on each plot. The average of a sample of individual plot sizes is then expanded to a per unit basis and the density compared.

The smallest plot that is the closest to the density of the largest plot is the most efficient. A similar procedure can be used to develop a species area curve for which the number of species is counted on different plot sizes until no more new species are added, and this plot size is the one selected. For habitat inventory, a 0.10-ac circular plot is adequate for trees, as is a 0.01-ac plot for shrubs (Table 4.10). A 9.6-ft² plot is a common size for sampling herbaceous vegetation because dry weight of vegetation in grams converts directly to pounds per acre. Otherwise, a plot 20 × 50 centimeters, divided into squares for estimates of percentage cover, is a frequently used configuration. Once the sizes of plots have been selected, the next problem is

TABLE 4.10
Circular Plot Sizes for Sampling

Size of Plot	Radius	
Metric Units		
0.10 m²	0.17 m	
1.00 m²	0.56 m	
10.00 m²	(0.001 ha)	1.78 m
100.00 m²	(0.01 ha)	5.64 m
1000.00 m²	(0.10 ha)	17.84 m
English Units		
0.001 ac	3 ft, 8.7 in.	
0.01 ac	11 ft, 9.3 in.	
0.025 ac	18 ft, 7.4 in.	
0.05 ac	26 ft, 4.0 in.	
0.10 ac	37 ft, 2.8 in.	
0.25 ac	58 ft, 10.5 in.	

9.60 ft² (20.97 in. radius): 1 g/9.60 ft² = 10 lbs/ac[a]

[a] Herbaceous vegetation.

to determine how many plots will be needed for a statistically valid inventory. The easiest approach is to make a preinventory sample and apply a variance formula to estimate numbers (Snedecor 1956).

$$N = (S^2 \times t^2)/d^2$$

where N is the number of samples required, S is standard deviation, t is the confidence interval at given degrees of freedom, and d is the margin of error (arithmetic mean times designated accuracy).

For example, when counting browse plants on a presample of 10 plots with computed $S = 6$, $t = 2.262$ (9 degrees of freedom at 0.95 confidence), and $d = 1.5$ (mean = 15, accuracy is designated at 0.10), then

$$N = (36 \times 5.11)/2.25) = 81.76 = 82$$

The number of sample plots needed as computed from the presample data is 82. There is a restriction in using the formula that the data must follow a normal distribution.

The number of sample points depends entirely on the variance of the item being sampled, but in many cases the amount of money available to do the job will be the deciding factor. Statistical techniques based on the cost of establishing plots as a factor in determining how many to establish are available. However, as a general rule a sample size from 2% to 5% of the area is a minimum.

4.10.3 SAMPLING DESIGN

Random samples are the best of all possible choices because they eliminate bias. However, such sampling is often difficult to do in the field because of problems in transferring a sample drawn from locations on a map to finding this location on the ground. To select a random sample for inventory, a dot grid is placed over the standard map or on aerial photographs, and each dot then becomes a sample point. Numbers are then drawn from a table of random numbers to select the points to sample. A *systematic sample* is more convenient to execute in the field, and most statisticians accept it with constraints, such as adding a random start component. The general procedure for a systematic sample is to randomly select starting points along the border of the area to be sampled and then locate the circular plots systematically at a given distance along parallel transects.

One of the easiest ways to reduce variation in any sample scheme is to *stratify* the area into like units. For example, if two vegetation types are present in an area, statistics can be calculated for each type separately rather than including both as one set of sample data. Stratification of vegetation types can be done with the aid of aerial photographs; the types are delineated on overlays for selecting the sample. Strata can be identified for many types of physical measurements, including vegetation types, physical size of trees in stands, administrative units, and so on. Whatever the criteria, the strata to be sampled should (Lund et al. 1978)

- Logically be related to the information that managers want sampled
- Exist in nature or have been artificially determined

- Represent a homogeneous condition that can be defined and is easily recognizable
- Have a definition that conforms to accepted standards

4.10.4 Information to be Collected

There have been many attempts to collect wildlife data in multiresource inventories in the past, but the rate of success has been variable. Estimating the density of animals, including amphibians, birds, mammals, and reptiles, cannot in most cases be done on plots established for a multiresource inventory because each animal group possesses unique behavioral traits requiring different sampling procedures. An integrated, multipurpose approach works well when the objective is to collect habitat information applicable to many species, such as the number of dead trees, stumps, and rotten logs used for cover.

The information to be collected in most inventories can generally be grouped into one of four categories: vegetation, physical features, signs of wildlife occupancy, and wildlife observations. These categories do not contain the entire domain of items to be listed or sampled but are intended as a preliminary list from which to begin selection. Information for specific areas can be added to it.

1. Vegetation
   - Dominant overstory species, basal area, percentage crown cover, diameter at breast height (dbh), trees per acre, and degree of grouping. *Basal area* is the cross-sectional area of a tree measured at diameter breast height (4.5 ft). *Crown cover* is the area of crown of trees or shrubs measured as a percentage of a plot.
   - Understory measurements for shrubs, small trees, and herbaceous plants; browse plant count, including form and age class, or a measure of basal area or crown density.
   - Forage production estimate. Clip and weigh material from small plots. Separate by grass and forbs.
   - The number of vegetation layers present and a measure of the amount of live foliage in each layer.
2. Physical features
   - Rock areas
   - Cliff
   - Seeps and springs
   - Caves
   - Lakes, ponds, and streams
   - Rainfall patterns
   - Slope and aspect
   - Topographic features (hills, valleys, etc.)
   - Burned areas by date and acreage
   - Hunt or management units
   - Cultural features of roads, trails, buildings, campgrounds, and recreation areas

3. Signs of wildlife occupancy
   - Fecal material or other signs (tracks, feathers, etc.)
   - Tree nests (squirrels and birds)
   - Ground nests
   - Hollow trees
   - Rodent mounds and tunnels
   - Roost trees (wild turkey, pigeons)
   - Squirrel middens (caches)
   - Ground dens (rabbit, bear, badger, coyote)
   - Snags (cavity-nesting birds and mammals)
   - Elk wallows, deer rubs
4. Wildlife observations
   - Observe all species seen or heard
   - Monitor nests, water holes, and special areas using video cameras or time-lapse photography
   - Use radio telemetry to determine movement
   - Use aerial counts and photography from aircraft

4.10.5   Uses of Habitat Data

After habitat data have been collected, the next step is to consider how the information is to be analyzed and displayed. Thanks to the availability of computers, consideration should be given to using graphics software to develop histograms and other forms of frequency diagrams for visual presentation. However, probably the most useful way to transfer information is by maps. Habitat data collected on sample plots can be used to develop three types of maps: (1) food, (2) cover, and (3) water. These maps should show at least the following items:

1. Food map
   - Shrub areas by species and acreage. Form and age class of woody plants should be shown.
   - Isograms of forage production based on data collected from sample plots. An *isogram* is a line connecting points of equal quantity, similar to an elevation contour on a map.
   - Locations of big trees that are likely mast producers. *Mast* is all fruits, nuts, and berries.
   - Concentrations of plant species that produce berries.
   - Natural openings and wet meadow areas by size and forage production.
2. Cover map
   - Plant reproduction thickets by species and size of area.
   - Location of caves, den trees, snags, and rock outcrops.
   - Dominant overstory species by size of area. Information can be drawn from timber type maps.
   - Roost trees and special features such as bear dens.

3. Water map
   - All streams, both perennial and intermittent.
   - Location of current or potential pollution sources.
   - All springs, wells, and permanent water bodies.
   - Isograms of the zones of influence (distances) around all water sources. Water-deficient areas will become apparent.
   - Important amphibian habitats (moist areas, seeps, and springs).

These maps can be made as overlays on aerial photographs or existing topographic maps of suitable scale, or even better, the data can be digitized and incorporated into a GIS and DBMS.

4.10.6   ESTIMATING ANIMAL NUMBERS

Without doubt, the most difficult job of any wildlife manager is to determine the density or total number of animals on a given unit of land. Until the details of an animal population are accurately known, the management of a unit lacks any clear direction. Numbers or indications of abundance are needed to evaluate carrying capacity or to determine the allowable harvest. A scale factor controls the estimation of animal populations just as it controls the making of habitat inventory maps. Recall that *scale* refers to the degree of resolution needed to meet management objectives. For example, a low-intensity habitat inventory generally requires only the same intensity of sampling for animals. The same relationship is true of high-intensity surveys. Clearly, there are two general ways of estimating animal numbers: counting animals or their sign.

4.10.7   COUNTING ANIMALS (DIRECT)

Direct methods depend on making a true *census,* which is a count of all the animals in an area or an estimate of the number by sampling. A manager seldom has the time and personnel to do a complete census. One direct method that can be used when staff are available is an animal drive, particularly if the area is small. Animals that depend on a territory and can be easily seen or heard (e.g., birds) can be censused by mapping a defended area. A census is approximated in a capture-recapture technique by trapping animals over 2 or 3 weeks until no more unmarked animals are recorded; the accumulated total is the animal population. Aerial counts work well for large animals such as deer, elk, and antelope when in open habitats or for species that flock. Counts for flocks should be augmented by aerial photographs, from which animal numbers can later be checked against observations.

Another direct way to estimate animal numbers is to use sample plots and mark-recapture methods. With the widespread availability of computers, many versions of mark-recapture formulas have evolved but are beyond the scope of this book. Many of the formulas for estimating animal numbers from mark-recapture methods are based on the equation known as the *Peterson* or *Lincoln index.* The Lincoln index is a good place to start when planning a survey that requires estimating populations because it is a simple ratio of marked to unmarked animals in a given time. The formula for the Lincoln Index is

$$M \div N = m \div n \text{ or } N = nM \div m$$

where M is the number marked during the first trapping period, m is the number of marked animals recaptured in the second trapping period, n is the number of animals captured in the second trapping period, and N is the population estimate.

For example, 35 squirrels were trapped in the first trapping period and 20 marked and 5 unmarked (25 total) in the second trapping period. The population estimate for these data is

$$N = 25 \times 35 \div 20 = 44$$

Confidence limits can be calculated for N using the formula

$$S_E = N = \sqrt{(N - M)(N - n)} \div Mn(N - 1)$$

where S_E is the standard error.

For a 95% confidence limit, two standard errors are added and subtracted from the population such that

$$S_E = 44 \sqrt{[(44 - 35)(44 - 25) \div 35(25)(44 - 1)]} = 2.96$$

$$\text{Upper limit} = 44 + 2(2.96) = 50$$

$$\text{Lower limit} = 44 - 2(2.96) = 38$$

All population estimation methods have advantages, disadvantages, and assumptions that must be considered when selecting the method best suited to the objectives of the survey. Direct methods are only used when there is a need for a high level of resolution, such as for setting hunting seasons or for associating numbers with habitat factors for research purposes. Estimating numbers of wildlife using direct techniques is seldom done except for certain game and threatened and endangered species. The reasons are that the techniques for arriving at a population estimate are time consuming and expensive, and the accuracy of the data is often not at a level that provides confidence in decision making. Care must be observed when discussing census figures or population estimates not to convey an accuracy that is not there. This is easy to do after time has been spent analyzing data with some of the sophisticated statistical formulas that exist for this purpose.

4.10.8 Counting Signs (Indirect)

Indirect methods rely on indexes developed from animal signs, such as tracks, fecal droppings, nests, or calls. One common technique is fecal (pellet) counts for deer as they commonly defecate pellets a fixed number of times a day. The number of droppings in sample plots is rapidly converted into deer per unit area or deer days use per unit area for a given number of days, ranging from 365 for yearlong use to as few as several weeks if the time of residence can be observed. Table 4.11 illustrates the pellet count (13 pellet groups per day) technique for mule deer. In addition to signs left by an animal, other indirect techniques depend on a frequency of observations to yield a population assessment as abundant, common, uncommon, or rare (Table 4.12).

TABLE 4.11
Pellet Count Technique for Estimating Deer and Elk Populations

Number of sample plots = 400
Size of sample plots   = 0.01 ac
Days of herd occupancy = 365 (time between counts)
Pellet groups counted   = 500
Pellet groups/plot   = 500/400 = 1.25

1. Sample area = No. plots × Size = 400 × 0.01 = 4 ac

2. Groups/ac/yr = Groups counted ÷ Sample area
$$= 500 \text{ groups} \div 4 \text{ ac} = 124$$

3. Deer/ac/yr   = Groups/ac/yr ÷ Pellet groups/day × Time
$$= 124 \div (13 \times 365) = 0.0261$$

4. Deer/mi^2/yr = Deer/ac/yr × 640 = 0.0261 × 640 = 16.7

TABLE 4.12
Examples of Statements to Determine Relative Abundance of Mammals

Large Mammals

A = Very likely to be seen, by a trained observer, in large numbers in normal habitat in the appropriate season.

C = May be seen most of the time, in smaller numbers, in normal habitat in the appropriate season.

U = May be seen infrequently in small numbers in normal habitat in the appropriate season.

R = Occupies only a small percentage of its normal habitat or occupies a very specific limited habitat.

Small Mammals

A = Frequently observed or collected in numbers.

C = Often observed or collected.

U = Infrequently seen or collected, present in low numbers.

R = Seldom seen or collected, present in very low numbers.

Note: **Normal habitat is where an animal is expected to occur**: A, abundant; C, common; U, uncommon; R, rare.

TABLE 4.13
Abundance of Animals Based on Frequency of Occurrence

Abundance	Frequency Sightings/Efforts	Probability %
Very abundant	8/10	>80
Abundant	6–8/10	61–80
Common	4–6/10	41–60
Uncommon	2–4/10	21–40
Rare	1–2/10	1–20
Very rare	<1/10	<1

Frequency classes can also be used to approximate abundance (Table 4.13). In this method, stations are established to observe the presence or absence of animals for a given number of times throughout the year or by season. If an animal is observed 7 times in a sample of 10 observations, the probability is 70% of seeing an animal in that habitat and in similar conditions. This type of indirect abundance estimating depends on an observer who knows the habitat requirements of an animal and its behavioral characteristics in using the habitat (time of day, season of year, etc.). Frequency counts to estimate relative abundance are best suited for low-intensity surveys not associated with setting hunting regulations.

Signs of animals can be important in establishing what is in an area if it has never before been inventoried. Reconnaissance inventories are often made using plots to record presence or absence information to establish a base of relative abundance, which can in itself be used as an index to compare between years or as a tool in designing more detailed inventory techniques.

4.10.9 Tree Measurements

Trees that are recorded on inventory plots have certain characteristics in both size and condition. Size characteristics are the diameter in inches and total height in feet partitioned into crown and trunk (Figure 4.10). Diameter is used to construct stand stables that show density of trees by diameter classes that are mostly in 2-in. gradations (2, 4, 6, 8, 10, etc.). However, for describing tree cover for wildlife, the class spread can be increased into four or five classes, such as 1–4 in. (seedling-saplings), 4–10 in. (young), 10–17 in. (intermediate), and more than 17 in. (mature).

A small number of size classes makes it easier to identify use and increases the accuracy because it is doubtful that small differences of 2 or 3 in. contribute to the value of food and cover areas. The age classes listed are generic and not related specifically to any one of the 20 forest types. If a larger number of age classes is needed in local areas for a specific forest type or animal, then they should be used because size class can always be aggregated into a smaller number of age classes. In addition, a larger tree size class may need to be added. The use of size and age when referring

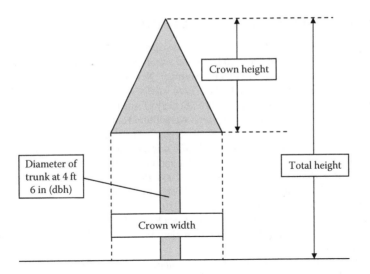

FIGURE 4.10   Tree measurements for describing forest components.

to trees is considered synonymous because of the strong linear relationship between the two.

Diameter (d) of trees is usually converted to basal area (BA), which measures the cross-sectional area at breast height (4.5-ft dbh). BA is calculated by the formula for the area of a circle. When the diameter is in inches and BA is in square feet, the formula is

$$BA = 0.005454154 \, d^2$$

For example, a tree with a 10-in. dbh has a BA of 0.54 ft². When the diameter is in centimeters and BA is in square meters, the formula is

$$BA = 0.000078539816 \, d^2$$

A tree with a diameter of 25.4 cm (10 in.) has a BA of 0.0507 m². BA is summed for a plot and converted to square feet or square meters per acre or hectare. BA is a measure of stand density, and biologists need to be familiar with its use by foresters and its meaning for wildlife. Diameter is easy to measure with a tape graduated so that diameter at breast height can be read directly from the circumference.

In addition to BA, another measure of stand density (trees per acre) is tree canopy cover. Canopy cover can be estimated from aerial photographs using stand density scales based on tree crown sizes or from plots in which crowns can be projected to the ground for measurement (Figure 4.10). Tree density, size, and crown area occupy a three-dimensional volume of space that wildlife can use for food and cover. A major problem in forest wildlife relationships is how to measure and describe this volume of space so that it can be meaningful in a decision-making process.

KNOWLEDGE ENHANCEMENT READING

Avery, T. E., and H. E. Burkhart. 2002. *Forest measurements.* McGraw-Hill, New York.

Burnham, K., and D. R. Anderson. 2002. *Model selection and multimodel inference: A practical information-theoretic approach.* Springer, New York.

DeStefano, S., and R. G. Haight (Eds.). 2002. *Forest wildlife-habitat relationships: Population community responses to forest management.* Society of American Foresters, Bethesda, MD.

Freese, F. 1962. *Elementary forest sampling.* Agriculture Handbook 232. U.S. Forest Service, Washington, DC.

Fulbright, T. E., and D. G. Hewitt. 2008. *Wildlife science: Linking ecological theory and management applications.* CRC Press, Taylor and Francis Group, Boca Raton, FL.

Gibbs, J. P., M. J. Hunter Jr., and E. L. Sterling. 2008. *Problem-solving in conservation biology and wildlife management.* Blackwell, Malden, MA.

McComb, B., B. Zuckerbury, D. Vesley et al. 2010. *Monitoring animal populations and their habitats.* CRC Press/Taylor & Francis, Boca Raton, FL.

Morrison, M. L. 2002. *Wildlife restoration.* Society for Ecological Restoration. Island Press, Washington, DC.

Skalski, J. R., and D. S. Robson. 1992. *Techniques for wildlife investigations: Design and analysis of capture data.* Academic Press, San Diego, CA.

Starfield, A. M., and A. L. Bleloch. 1986. *Building models for conservation and wildlife management.* Macmillan, New York.

Walters, C. 1986. *Adaptive management of renewable resources.* Macmillan, New York.

REFERENCES

Beier, P. 1993. Cougar dispersal in fragmented habitat. *J. Wildl. Manage.* 40(4):434–440.

Beier, P., and S. Loe. 1992. A checklist for evaluating impacts to wildlife movement corridors. *Wildl. Soc. Bull.* 20:434–440.

Billings, W. D. 1970. *Plants, man, and the ecosystem.* Wadsworth, Belmont, CA.

Brown, E. R. (Ed.). 1985. *Management of wildlife and fish habitats in forests of western Oregon and Washington.* Part I. U.S. Forest Service, Pacific Northwest Region, in cooperation with U.S. Bureau of Land Management, Portland, OR.

Bunnell, F. L. 1989. *Alchemy and uncertainty: What good are models?* General Technical Report PNW-GTR-232. U.S. Forest Service, Portland, OR.

Christensen, A. G., L. J. Lyon, and J. W. Unsworth. 1993. *Elk management in the northern region: Considerations in forest plan updates or revision.* General Technical Report INT-303. U.S. Forest Service, Ogden, UT.

Clements, F. E. 1916. *Plant succession: An analysis of the development of vegetation.* Carnegie Institute, Publication 242. Washington, DC.

Clements, F. E. 1949. *Dynamics of vegetation.* Hafner, New York.

Cole, G. F. 1958. *Range survey guide.* Federal Aid Project W-37-R. Montana Department of Game and Fish, Missoula.

Commission for Environmental Conservation. 1997. *Ecological regions of North America. Toward a common perspective.* Commission for Environmental Conservation, Montreal.

Crance, J. H. 1987. *Guidelines for using the Delphi technique to develop habitat suitability index curves.* Biology Report 82. U.S. Fish and Wildlife Service, National Ecology Center, Fort Collins, CO.

Dasmann, R. F. 1951. Some deer range survey methods. *Calif. Fish Game* 37:43–52.

Dasmann, W. P. 1948. A critical review of range survey methods and their application to deer range management. *Calif. Fish Game* 34:189–207.

DeSteffano, S., and R. C. Haight (Eds.). 2002. *Forest wildlife-habitat relationships: Population and community responses to forest management.* Society of American Foresters, Bethesda, MD.

Drew, T. J., and J. W. Flewelling. 1977. Some recent theories in yield-density relationships and their application to Monterey pine plantations. *Forest Sci.* 23:517–534.

Einarsen, A. S. 1948. *The pronghorn antelope and its management.* Wildlife Management Institute, Washington, DC.

Elton, C. E. 1966. *The pattern of animal communities.* Methuen, New York.

Galli, A. E., C. F. Leck, and R. T. T. Forman. 1976. Avian distribution patterns in forest islands of different sizes in New Jersey. *Auk* 93:356–364.

Giles, R. H., Jr., and N. Snyder. 1970. Simulation techniques in wildlife habitat management. In D. Jameson (Ed.). *Modeling and systems analysis in range science,* Series Number 5, pp. 23–49. Range Science Department, Colorado State University, Fort Collins.

Grubb, T. G. 1988. *Pattern recognition: A simple model for evaluating wildlife habitat.* Research Note RM-487, U.S. Forest Service, Fort Collins, CO.

Haapanen, A. 1965. Bird fauna of the Finnish forests in relation to forest succession. I. *Annu. Zoo. Fenn.* 2:153–196.

Harris, L. D. 1984. *The fragmented forest: Island biogeography, theory and the reservation of biological diversity.* University of Chicago Press, Chicago.

Hewitt, O. H. (Ed.). 1967. *The wild turkey and its management.* Wildlife Society, Washington, DC.

Holbrook, H. L. 1974. A system for wildlife habitat management on southern national forests. *Wildl. Soc. Bull.* 2:119–123.

Holling, C. S. (Ed.). 1978. *Adaptive environmental assessment and management.* Wiley International Series on Applied Systems Analysis. Vol. 3. John Wiley and Sons, Chichester, UK.

Hoover, R. L., and D. L. Wills (Eds.). 1984. *Managing forested lands for wildlife.* Colorado Division of Wildlife in cooperation with U.S. Forest Service, Rocky Mountain Region, Denver, CO.

Jeffers, J. N. R. 1980. *Modeling, statistical checklist.* Institute of Terrestrial Ecology, Natural Environment Research Council, Cambridge, UK.

Lund, H. G., V. J. LeBau, P. F. Ffolliott, W. W. Robinson. 1978. *Integrated inventories of renewable natural resources.* General Technical Report RM-55. U.S. Forest Service, Fort Collins, CO.

MacArthur, R. H., and E. O. Wilson. 1967. *The theory of island biogeography.* Princeton University Press, Princeton, NJ.

MacClintock, L. R., R. Whitcomb, and B. Whitcomb. 1977. Evidence for the value of corridors and minimization of isolation in preservation of biotic diversity. *Am. Birds* 31:6–16.

Mannan, R. W., M. L. Morrison, and E. C. Meslow. 1984. The use of guilds in forest bird management. *Wildl. Soc. Bull.* 12:426–430.

Marcot, B. G. 1986. Concepts of risk analysis as applied to viable population assessment and planning. In B. A. Wilcox, P. F. Brussard, and B. G. Marcot (Eds.), *The management of viability: Theory, applications and case studies,* pp. 89–101. Center for Conservation Biology, Department of Biological Science, Stanford University, Stanford, CA.

Mason, R. D. 1970. *Statistical techniques in business and economics.* Richard D. Irwin, Homewood, IL.

McTague, J. R., and D. R. Patton. 1989. Stand density index and its application in describing wildlife habitat cover. *Wildl. Soc. Bull.* 17:58–62.

Mitchell, T. R., and B. M. Kent. 1987. Characterization of the FORPLAN analysis system. In T. W. Hoekstra, A. A. Dyer, and D. C. Le Master (Tech. Eds.), *FORPLAN: An evaluation of a forest planning tool,* pp. 3–14. General Technical Report RM-140. U.S. Forest Service, Fort Collins, CO.

Mollohan, C. 1988. *Merriam's turkey expert opinion survey.* Arizona Game and Fish Department Report. State of Arizona, Phoenix, AZ.

Morrison, M. L., B. G. Marcot, and R. W. Mannan. 1998. *Wildlife-habitat relationships, concepts and applications.* University of Wisconsin Press, Madison.

Mosby, H. S. (Ed.). 1963. *Wildlife investigational techniques.* Wildlife Society, Washington, DC.

Nelson, R. D., and H. Salwasser. 1982. The Forest Service wildlife and fish habitat relationships program. *Trans. N. Am. Wildl. Nat. Resource Conf.* 47:174–182.

New Mexico Game and Fish Dept. 1973. Big game habitat analysis for New Mexico [mimeo]. N.M. Game and Fish Department, Santa Fe.

Patton, D. R. 1975. A diversity index for quantifying habitat "edge." *Wildl. Soc. Bull.* 3:171–173.

Patton, D. R. 1978. *Run wild: A storage and retrieval system for wildlife habitat information.* General Technical Report RM-51. U.S. Forest Service, Fort Collins, CO.

Patton, D. R. 1984. A model to evaluate Abert squirrel habitat in unevenaged ponderosa pine. *Wildl. Soc. Bull.* 12:409–414.

Patton, D. R. 1987. Is the use of management indicator species feasible. *West. J. Appl. Forest.* 2:33–34.

Patton, D. R. 1993. Toward multispecies management across landscapes. *J. Forest.* 91(7):7.

Patton, D. R., and J. M. Hall. 1966. Evaluating key areas by browse age and form class. *J. Wildl. Manage.* 30:476–480.

Phillips, E. A. 1959. *Methods of vegetation study.* Henry Holt and Company, New York.

Pimlot, D. H. 1969. The value of diversity. *Trans. N. Am. Wildl. Nat. Resource Conf.* 34:265–280.

Reineke, L. H. 1933. Perfecting a stand density index for even-aged forests. *J. Agric. Res.* 46:627–638.

Reynolds, R. T., R. T. Graham, M. H. Reiser, et al. 1991. *Management recommendations for the northern goshawk in the southwestern United States.* U.S. Forest Service, Southwestern Region, Albuquerque, NM.

Rogers, J. L., J. M. Prosser, and L. D. Garrett. 1981. Modeling on-site multiresources effects of silviculture management prescriptions. In *Forest management planning: Present practice and future decisions, Proceedings: IUFRO Symposium,* Publication FWS-1-81, pp. 107–117. School of Forestry and Wildlife Resources, Virginia Polytechnic Institute and State University, Blacksburg.

Root, R. B. 1967. The niche exploitation pattern of the blue-gray gnatcatcher. *Ecol. Monogr.* 37:317–349.

Salwasser, H., C. K. Hamilton, W. B. Krohn, et al. 1983. Monitoring wildlife and fish, mandates and their implications. *Trans. N. Am. Wildl. Nat. Resource Conf.* 48:297–307.

Schantz, H. L. 1911. *Natural vegetation as an indicator of the capabilities of land crop production in the Great Plains Area.* Bulletin 210. Bureau of Plant Industry, U.S. Department of Agriculture, Washington, DC.

Scott, J. M., F. Davis, B. Csuti, et al. 1993. *Gap analysis: A geographic approach to protection of biological diversity.* Wildlife Monograph 123. Wildlife Society, Washington, DC.

Scott, V. E., and D. R. Patton. 1989. *Cavity-nesting birds of Arizona and New Mexico forests.* General Technical Report RM-10. U.S. Forest Service, Fort Collins, CO.

Severinghaus, W. D. 1981. Guild theory development as a mechanism for assessing environmental impacts. *Envir. Manage.* 5:187–190.

Shafer, E. L. 1963. The twig-count method for measuring hardwood deer browse. *J. Wildl. Manage.* 27:428–437.

Short, H. L., and K. P. Burnham. 1982. *Technique for structuring wildlife guilds to evaluate impacts on wildlife communities.* Special Science Report on Wildlife 244. U.S. Fish and Wildlife Service, Fort Collins, CO.

Smith, F. W., and J. N. Long. 1987. Elk hiding and thermal cover guidelines in the context of lodgepole pine stand density. *West. J. Appl. Forest.* 2:6–10.

Snedecor, G. W. 1956. *Statistical methods.* Iowa State Press, Ames.

Starfield, A. M., and A. L. Bleloch. 1986. *Building models for conservation and wildlife management.* Macmillan, New York.

Steger, T. D. 1986. *Topographic maps.* U.S. Geological Survey, Washington, DC.

Stickney, P. F. 1966. Browse utilization based on percentage of twig numbers browsed. *J. Wildl. Manage.* 30:204–206.

Szaro, R. C. 1985. Guild theory: Fact or fiction. *Environ. Manage.* 10:681–688.

Taber, W. D., and R. P. Dasmann. 1958. *Black-tailed deer of the chaparral.* Game Bulletin Number 8, California Department of Game and Fish, Sacramento.

Talbott, M. W. 1937. *Indicators of southwestern range conditions.* Farmers Bulletin 1782, U.S. Department of Agriculture, Washington, DC.

Taylor, W. P. 1940. Ecological classification of the mammals and birds of Walker County, Texas, and some adjoining areas. *Trans. N. Am. Wildl. Conf.* 5:170–176.

Thomas, J. W. (Ed.). 1979a. *Wildlife habitats in managed forests: The Blue Mountains of Oregon and Washington.* Agriculture Handbook 553. U.S. Department of Agriculture, Washington, DC.

Thomas, J. W., R. J. Miller, C. Maser, et al. 1979b. Plant communities and successional stages. In J. W. Thomas (Ed.), *Wildlife habitats in managed forests: The Blue Mountains of Oregon and Washington*, pp. 22–23. Agriculture Handbook 553. U.S. Forest Service, Washington, DC.

Thomas, J. W., C. Maser, and J. E. Rodiek. 1979c. Riparian zones. In J. W. Thomas (Ed.), *Wildlife habitats in managed forests: The Blue Mountains of Oregon and Washington*, pp. 40–47. Agriculture Handbook 553. U.S. Forest Service, Washington, DC.

Thomas, J. W., and D. E. Toweill (Eds.). 1982. *Elk of North America: Ecology and management.* Wildlife Management Institute. Stackpole Books, Harrisburg, PA.

U.S. Congress. 1934. Fish and Wildlife Coordination Act. Public Law 85-624. U.S. Government Printing Office, Washington, DC.

U.S. Congress. 1960. Multiple-Use, Sustained Yield Act. Public Law 86-517. U.S. Government Printing Office, Washington, DC.

U.S. Congress. 1967. Sikes Act. Public Law 93-452. U.S. Government Printing Office, Washington, DC.

U.S. Congress. 1969. National Environmental Policy Act. Public Law 91-190. U.S. Government Printing Office, Washington, DC.

U.S. Congress. 1973. Endangered Species Act. Public Law 93-205. U.S. Government Printing Office, Washington, DC.

U.S. Congress. 1974. Forest and Rangeland Renewable Resources Planning Act. Public Law 93-378. U.S. Government Printing Office, Washington, DC.

U.S. Congress. 1976. National Forest Management Act. Public Law 94-588. U.S. Government Printing Office, Washington, DC.

U.S. Department of Agriculture (USDA). 1974. *Wildlife habitat evaluation program (WHEP).* U.S. Forest Service, Southeastern Area, State and Private Forestry, Region 8, Atlanta, GA.

U.S. Department of Agriculture (USDA). 1981. *An assessment of the forest and rangeland situation in the United States.* U.S. Forest Service, Washington, DC.

U.S. Department of Agriculture (USDA). 1982. *Policy on fish and wildlife.* Secretary's Memo 9500-3. Office of the Secretary of Agriculture, Washington, DC.

U.S. Department of Agriculture (USDA). 1984. *Wildlife, fish, and sensitive plant habitat management.* Forest Service Handbook 2600, Amendment 48. U.S. Government Printing Office, Washington, DC.

U.S. Department of Agriculture (USDA). 1988. *Integrated resource management.* U.S. Forest Service, Southwestern Region, Albuquerque, NM.

U.S. Department of the Interior (USDI). 1980. *Habitat as a basis for environmental assessment*. Ecosystem Management 101, Release 4-80. Division of Ecol. Serv., U.S. Fish and Wildlife Service, Washington, DC.

Verner, J. 1983. An integrated system for monitoring wildlife on the Sierra National Forest. *Trans. N. Am. Wildl. Nat. Resources Conf.* 48:355–366.

Verner, J. 1984. The guild concept applied to management of bird populations. *Envir. Manage.* 8:1–14.

Verner, J., and A. Boss. 1980. *California wildlife and their habitats: Western Sierra Nevada*. General Technical Report PSW-37. U.S. Forest Service, Berkeley, CA.

Verner, J., M. L. Morrison, and C. J. Ralph. 1984. *Wildlife 2000, modeling habitat relationships of terrestrial vertebrates*. University of Wisconsin Press, Madison.

Wallmo, O. C. 1981. *Mule and black-tailed deer of North America*. A Wildlife Management Institute Book in cooperation with U.S. Forest Service. University of Nebraska Press, Lincoln.

Wallmo, O. C., L. H. Carpenter, W. L. Regelin, et al. 1977. Evaluation of deer habitat on a nutritional basis. *J. Range Manage.* 30:122–127.

Walters, C. 1986. *Adaptive management of renewable resources*. Macmillan, New York.

Williams, G. L., K. R. Russell, and W. K. Seitz. 1977. Pattern recognition as a tool in the ecological analysis of habitat. In A. Marmelson (Chair), *Proceedings: Classification, inventory, and analysis of fish and wildlife habitat*, pp. 521–531. U.S. Fish and Wildlife Service OBS, Washington, DC.

Yeager, L. E. 1961. Classification of North American mammals and birds according to forest habitat preference. *J. Forest.* 59:671–674.

5 Forest Attributes and Wildlife Needs (FAAWN)

5.1 BACKGROUND INFORMATION

The structure and layers in a forest (Table 1.1) form the foundation of a data model design (Figure 5.1) using relational theory to identify physical components (Table 1.2) that are used for food and cover areas by animal species. The FAAWN (forest attributes and wildlife needs) data model in its current form was developed as an educational tool, but it can also be used for application in the work environment. The intent of the model is to show relations between forest components and animals that occur in forested areas. It is a design that can be used to explore combinations of attributes. In the data model scheme, there is a chain of descriptive attributes in columns and a web of information in the relations of linking tables. Data in the tables and columns can be manipulated or changed without losing information in the individual data cells.

Manipulating data is accomplished by selecting, combining, and filtering through a menu system or by developing queries using structured query language (SQL) statements developed for relational databases such as Microsoft Access. The data model contains considerable information in most of the tables listed in Figure 5.1 with the exception of the Life-Needs (food and cover components) and the Hypothesized (animals in types in states) tables. The Life-Needs table has information for 411 animals in the data cells, but the Volume column cells have only a few entries. Also, the Life-Needs table contains animals other than those considered as forest species. These species are examples of animals associated with other land forms (grassland, wetland, shrubland, etc.). The Hypothesized table (identified as number 14 in Figure 5.1) contains comprehensive information from my original data in the RUNWILD and WILDHARE data models that have been used in Arizona and New Mexico but not in other states. However, the references suggested at the end of Chapter 5 do contain some of the tree and canopy size and density data that can be associated with animals in different parts of the United States.

Because FAAWN is intended primarily as an educational tool, the many ways to manipulate the data are left for discovery by the user except for some basic "getting started skills." As a result, there are no predetermined queries provided so that to learn the detail contained in the data model a user will need to learn skills that ultimately will result in a better understanding of the chains and webs that exist in a forest ecosystem. FAAWN is provided without copyright restrictions. The only requirement is that a user reference *Forest Wildlife Ecology* as the source of information.

The desired outcome of the data model is to provide the basic structure for a transportable scheme that can account for all forest components and how they can be

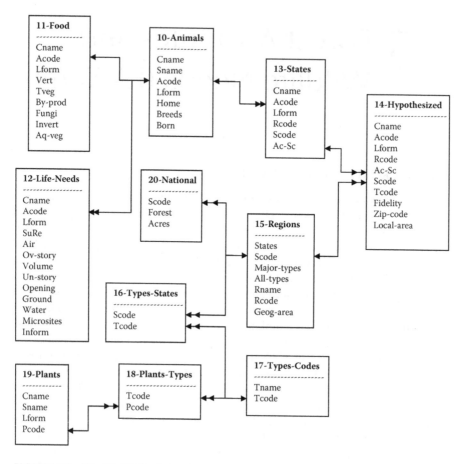

FIGURE 5.1   The FAAWN relational data model design with table names and column headings. The data model design also can be printed from within the FAAWN data model.

associated with different animals or groups of animals. It is easy to use the data model to filter and sort information in the relations (tables) and attributes (columns) that can be developed for a local database for analysis and application. Data in FAAWN can be used as a starting point to be transported to other areas depending on its validation for species for a type in a state and specific locality. The structure without data can be copied into a new database to populate with local information in developing a new relationships data model. The Hypothesized table indicates the following:

- Animals in the table have been identified from distribution maps as being in a state.
- Forest types (Tcodes) in the table have been identified from the National Atlas as being in a state.
- Approximately 550 of the 1,578 animals listed in the Animals table have been identified as potential forest type associates.

- It is reasonable to expect a high proportion of animals that are forest associates in a state can be found in forest types in that state.
- By combining the animals in a state with types in a state, there are over 63,000 listings of animals in forest types in states.
- Animals can occur in a type in one state but not in the same type in another state, and for this reason the listing of animals by state in types is hypothetical until they are validated for the state and type.

The following sections and paragraphs contain information in a structured format that is not always easy to understand because relational database development has strict guidelines that must be followed. The data model described in this chapter was developed using protocols for relational databases, but experience has shown that some deviations from rules is often necessary and beneficial for users. In the following sections, each table is explained and, if appropriate, information is provided on how the columns are defined and if options exist on how data can be formatted and displayed. A data model is also a synonym for the term *database scheme*, so the two can be used interchangeably.

Chapter 5 is presented in a way to induce thinking about how forests have components that provide food and cover areas for many species of wildlife. Information in FAAWN is linked similar to an ecological web, so that when a selection is made for a specific component, then all species that potentially could use the same component are identified. It is important to read the following sections on database management systems to have a basic understanding of how relations and attributes make the system work efficiently and how relations can help biologists and foresters understand some very complicated forest-wildlife interactions.

One of the difficulties of reading and synthesizing hundreds and thousands of pieces of information from the research literature interpreted by a second person (data model developer) to a third party (a user) is in how data are presented so that all three have a common understanding or interpretation of what was intended. Many good codes are used in the FAAWN data model to identify or describe things. Some codes are stable with little or no change, and others are sometimes fuzzy and not so descriptive.

An example of a good code is the condensed version of a scientific name used as an identifier of a plant or animal. A "not-so-good code" is the number "1" to indicate that an animal is known to be present in a forest type (Table 5.2, Fidelity). This type of code is a generalization that is necessary but not too useful for management until it is strengthened with other information, such as adding "h" for habitat or "f" as fringe. These two codes are fuzzy, and one has to read between the lines to determine what they mean. Fuzziness can be eliminated by adding additional descriptors, such as time of year (w = winter, s = summer). Other descriptors can be added when needed.

The current commercial relational database software packages use tables, which are columns in a matrix of data cells with a linking scheme to store information, and there is no good way to entirely eliminate codes without making them more complex. However, pattern recognition and neural network programs are being developed that in the future will allow sentence structure to replace codes, but these programs are several years away. The best advice for looking at codes that are not well defined or are too general in nature is that if they do not seem to fit the issue, then do not use them or add your own descriptors.

5.2   RELATIONAL DATABASE SYSTEMS

5.2.1   Structure and Organization

Data for storage in a database must be structured and organized in a logical and consistent way. The fundamental structure is a record devoted to an identifiable entity (e.g., species). Within this record, a varying number of fields are located, each dedicated to holding information or a specific characteristic associated with the entity. Similar types of data must be stored in similar locations in the database. Thus, taxonomic data are not typically stored in a habitat database intended for management information. Irrelevant data take up space and reduce efficiency.

5.2.2   Quality Control

One of the best ways to control the quality of data entered into a database is to have a structured format entry form that is easily followed by a data entry operator. Until about 1980 almost all wildlife databases were organized in a hierarchical system in which categories such as animal species and states were stored in a table-and-column format without any links from one table to another. Because humans quite typically think, organize, and model our world in a hierarchical manner, this type of scheme is both a comfortable and an intuitive approach to data management. But, hierarchical databases have major drawbacks; without linkages, they are relatively inflexible and hard to change. Furthermore, they usually provide only one path to a particular record.

5.2.3   Relation Theory

Relational database theory was developed by Codd (1970). The microcomputer had not been developed, so the relational approach was first conceived for large mainframe computers. However, neither the type of computer nor the available software packages had the capacity to link data to show relationships in a fuller, more natural holistic way. With rapid advances in miniaturization and increases in memory and storage in the early 1980s, the electronic structure of microcomputers advanced rapidly and became as powerful as the earlier mainframes. This resulted in the larger and more complex systems developed for mainframes being transformed and often augmented for use in developing large relational databases on a microcomputer.

A *relational database* is a collection of relations and attributes stored together in a logical way.

While the body of data is stored in a file system, a number of such files are linked in a variety of ways not unlike ecological food chains and webs. The manner in which data are formatted and linked is largely defined by the many ways in which relational software packages have built on Codd's initial formulations. Software packages to manage relational databases for microcomputers having several hundreds of relations (tables) and attributes (columns) and tens of thousands of rows for data are now commercially available. The limitations of relational databases are controlled only by the size of the hard drive and central processing unit (CPU) speed.

5.2.4 RELATIONSHIP SYSTEMS

The term *relation* refers to an accumulation of facts (attributes) describing the same subject. An example of a subject might be students. The question is, What do students have in common? They have a first and last name, weight, date of birth, height, social security number, and so forth. All the characteristics pertaining only to students form a relation. The different characteristics just mentioned are attributes of each student. Therefore, an *attribute* is a fact about an inherent physical characteristic, property, or measurement of a particular subject.

In a database scheme, a relation is a table, and an attribute is a column in that table. The intersection of rows with columns creates a *data cell* containing information about the subject of the column. Some advantages of relational databases are as follows:

- They retain the simplicity and intellectually comfortable style associated with data in tabular form.
- Two relations can be used to develop a third relation for either permanent or temporary use.
- They have the ability to link and associate different relations.
- New relations can be added or deleted at any time, and no data are lost.

The utility of relational databases is found in the simplicity and control of data entered in the tables. If the tables are well organized in the system, any particular piece of data is entered only once. A relational database allows retrieval of information from linking tables so that the combined information provides more thorough answers to many questions.

A *relationships system* is information categorized, stored, and linked in a predesigned table-and-column format in a database.

The use of relational databases implicitly imparts a very particular way of thinking and conceptualizing information. The procedures followed to link information forces the user to take into account the way that the natural world is organized not only as hierarchical chains but also as webs in a complex of relationships (e.g., as in an ecosystem). Only in recent years has database software for storing and retrieving data conforming to relational theory become available for microcomputers. A relational data model is a concept based on mathematical set theory defining how users interpret data organization (Fleming and von Halle 1989).

As a concept, a data model is orderly, predictable, and intuitive and is used to perceive, organize, and manipulate data. The concept categories below are applicable and basic to wildlife in forest areas everywhere. Large amounts of information are typically available, but the important items to consider are those that will be used in decision making or to facilitate a reporting process. The general categories of information contain subject matter as follows:

1. Characteristics of wildlife species
   a. Scientific name
   b. Common name
   c. Legal status
   d. Life history information (includes food habits, home ranges, etc.)

2. Habitat characteristics used by wildlife
   a.   Physical description of feeding areas
   b.   Physical description of cover areas
3. Wildlife associated with vegetation types
   a.   Which wildlife species are associated with which vegetation types
   b.   Value of the vegetation to different species (habitat, fringe)

The usefulness of information in a table is greatly enhanced if the idea of a central theme is followed so that columns in the table are directly related to the theme.

5.2.5   Design Criteria

The three basic categories of information identified for inclusion in a relational database, along with individual items in each category, should be developed in consultation with field biologists to determine which subjects generally are used at the local level for decision making. Each of the three broad categories can be used as the subject of a table to define the contents of columns containing more detailed information for a matrix. My discussions with foresters and wildlife biologists over many years have led to six design criteria that a habitat data model should meet:

- Information in the model must be suitable for use in a table format using rows and columns.
- The model must be adaptable for storing in a microcomputer for use with relational database software.
- Information stored in the database must be suitable for decision making.
- The model must have the capability to answer nested questions.
- The model must have the capability of identifying animal species using the same or overlapping physical habitat factors.
- Data in the model must be displayed without redundancy.

These criteria can all be met if restrictions are enforced to describe habitat by relations and attributes conforming to relational theory. The problem is one of storing summary data in tables, columns, and rows for ease in understanding and for reporting purposes. Each data cell has a logical and physical domain. The *logical domain* corresponds to the subject of the data in a column. For example, data cells containing common names should not contain scientific names. The *physical domain* refers to the types of data: real, integer, date, time, text, and so on.

Data cells can have one item or several items in a group that can be repeated with a delimiter such as a comma. Ordinarily, data cells do not contain repeating groups, but this is a trade-off in design and how the information is to be visually presented. By following a set format, the data will be consistent and can be read easily from field forms for computer entry. Several characteristics are common to tables in relational databases:

- Every row has the same number of columns.
- Each column contains the same kind of fact (logical domain) in each row.
- There should be only one entry for each row for each fact.
- The order of the rows is not important.
- The order of the columns is not important.
- No two rows are exactly the same.

5.2.6 How Relational Systems Work

Relations have key attributes for retrieving data from more than one table simultaneously. Data in one relation are linked to other relations by direct and indirect routes. Direct links are provided by a unique identification code (key) linking every row in one table with every row in another table in which the codes are identical. Indirect links occur when one relation contains links to other relations that are not directly linked by the unique identification code.

One-to-One Relationships: In this case, one row in one table is linked only to one row in another table. The link between two tables is indicated by a solid line linking columns.

One-to-Many Relationships: One row in one table is linked to many rows in another table. To indicate the relationship in a diagram, a "1" or arrow (>) identifies the one side of the relation and two arrows (>>) or an infinity sign (∞) identifies the many side (Figure 5.1).

Many-to-Many Relationships: These relationships can be identified by many rows in one table linking many rows in another table. Many-to-many relationships are not as efficient as one-to-many relationships and need to be redesigned by including a linking table.

5.2.7 Rules for Placing the Links

In one-to-one relationships, the link should be a stable attribute, or it should be taken from the table that by definition contains a key column. In one-to-many relationships, the linking column should be whichever column is defined as the *key column*. Information from two tables can be queried if there is a link between tables. One of the safest and most certain methods of constraining entry is to write data entry rules into the input software for each table. Thus, the software governing the table input might contain a code that will elicit an error message if the operator enters data on a new item without first having entered the data in a core table. Software rules can also be written to prevent input of duplicate entries in all or only specified tables.

When making a query for information from more than one table, the relationships between tables are important because they determine how the information will be displayed when retrieved. The power and utility of the FAAWN model is in the combinations of attributes within tables associating animals with their use of forest components. Information can be projected into new tables that can be deleted after they have served their purpose. The accuracy of the data model and the relationships that can be inferred are only as good as the information that has been authenticated and entered in the attributes. If attribute information previously has been validated through experimentation, statistical analysis, or repeated field observations, then the relationships will be accurate.

One important feature of relational databases is the ability to make computations. For example, if acres is a column in a table, then the data in these rows and columns can be added by using the appropriate command. Averages, ranges, and so forth can be computed using similar commands. There is no type of habitat information that cannot be entered into relational databases for developing useful wildlife habitat relationships. As more information is added to the wildlife literature, and as management decisions become more complex, the use of increasingly powerful microcomputers using relational database software will provide wildlife biologists and foresters the ability to compare, select, and make decisions based on the essential facts contained in massive databanks.

5.2.8   DETERMINING THE LINKS

Determining the proper links and relationships between tables is as much an art as it is a science. Intuition plays an important role in determining the links between tables and the dependencies that are created. Small test databases well "worked over" will in time identify the anomalies that exist, and once these are known, a solution to problems can be found. An entire database should never be loaded into a data model without first making test runs.

Trade-offs must be made in database design. For example, one table may have to be split into two tables for logical reasons, but doing so may then lead to redundancy of data and more work for the data entry operator. Tables with many columns, however, generally are not efficient except in the case of a core table. The developer of the data model also has to recognize the capacity of a printer to print information on a given paper size. This restriction forces the use of codes for attributes to print the appropriate relationships on a single sheet of paper (e.g., a field data form or report size).

5.2.9   EVALUATE AND TEST THE DESIGN

The last step is to evaluate and test the design of the structure of the database. In this step, look for design flaws that might lead to incorrect or redundant data. Every table in the database must be evaluated by asking the following questions:

- Does the table have a central theme or subject?
- Is each column a physical or logical fact about the subject?
- Does the table have a key column?
- Is the database easy to use?

The goal of good database design is to include enough detail so that the only question that cannot be answered is the question that is never asked (Kroenke and Nilson 1986). In the final analysis, the only way to test the data model is to ask questions for which the outcome is known. At first, load just enough data into the database to answer such test questions. Test data can, in fact, be "dummy" information to make it easy to recognize a problem in the database when the answer is presented.

Validation and testing a data model containing research results, commonsense associations, and educated guesses will come from research data, repeated field observations, and continuously correcting errors (adaptive resource management). Attribute information, such as associating an animal with a vegetation type, is not difficult to validate, but the process is much more complicated when relating an animal to a stand structural stage at a given time of year. This type of validation requires several years of data because of the yearly changes in environmental conditions that can affect the use of habitat by a species (particularly birds).

5.3   DEVELOPING A RELATIONAL DATA MODEL

The model presented in this chapter resulted from the early work of many wildlife biologists in federal and state agencies to meet the needs of the Forest and Rangeland Renewable Resources Planning Act (U.S. Congress 1974). The act required a national assessment of natural resources in the United States, and the information used is still available today in many publications and databases. As a result of new relational software that uses SQL in a menu system, it is now easier to design an efficient relational scheme to manipulate large amounts of data for field application. The FAAWN data model incorporates suggestions from practicing field biologists and students along with new information from published literature.

The data model design (Figure 5.1) is a web of information that includes 11 relations (Table 5.1), but others can be added when needed. For the sections that follow, a directory of definitions and a menu for each column (attribute) is in Table 5.2. The directory lists in alphabetical order all column names for each table in Figure 5.1. In

---

TABLE 5.1
Relations in the FAAWN Data Model

Relations	Description
10-Animals[a]	Animal species in the United States
11-Food	Items in food groups eaten by animals
12-Life-Needs	Forest wildlife food and cover components
13-States	Forest animals in each state
14-Hypothesized	Forest animals in each type in each state
15-Regions	Summary of regions, states, and types
16-Types-States	A vertical list of forest types in each state
17-Types-Codes	Type names and codes
18-Plants-Types	A vertical list of types and plant codes
19-Plants	Plant names and codes by life form
20-National	National forests in each state

[a] The term *Animals* in this text is a synonym for wildlife.

TABLE 5.2
Directory for Tables (Relations) and Columns (Attributes) in Figure 5.1 with Menus for Data Cells

Acode. (10 spaces). Animal name code.

> For efficiency in manipulating data and to reduce errors, the data model has many abbreviations, and the first one of importance is a code for scientific name (Acode). Codes are used to link one table with another in a direct way.

> Acode is a primary key in a one-to-many relation. When a primary key is designated, then an entry on the many side cannot be made unless there is an Acode already entered on the one side that links to another table with an Acode.

> Acode is developed from Sname by using three letters from genus and three letters from species. If duplicates occur in the six-letter combination, then the last letter is replaced with a number (e.g., 1, 2, etc.). Example: Lepcal becomes Lepca1 or Lepca2, and so on to keep the six-letter combination.

> An Acode is entered into the Wildlife relation only once and is a link to common and scientific name. The attributes in Wildlife all have a direct association with an animal in the table so that if Acode is known then the other attributes listed are also known.

Acres. (15 spaces). Acres in a national forest.

Ac-Sc. (15 spaces). A code using Acode and Scode to link the States and Hypothesized tables.

Air. (10 spaces). Air space for feeding.
a = Feeds in air over/through canopy
b = Hunts from air
c = Feeds while hovering

All-types. (40 spaces). All the types that occur in a state.

Aq-veg. (20 spaces). Aquatic vegetation eaten by animals.
su = Submergents
em = Emergents
al = Algae
pl = Plankton
ai = Aquatic insects
go = General omnivore

> Use commas to separate codes; for example, in the "go" category, there could be "fi, sm, bi" as the entry describing fish, small mammal, or bird.

Born. (10 spaces). Time of year when young are born. Time can be over several months.
1 = January
2 = February
3 = March
4 = April
5 = May
6 = June
7 = July
8 = August

TABLE 5.2 (continued)
Directory for Tables (Relations) and Columns (Attributes) in Figure 5.1
with Menus for Data Cells

    9 = September
    10 = October
    11 = November
    12 = December
    13 = Any month

    Example: 10–12 is October to December.

Breeds. (10 spaces). The time of year breeding occurs. Time can be over several months.

    ja = January
    fe = February
    mr = March
    ap = April
    ma = May
    jn = June
    jl = July
    au = August
    se = September
    oc = October
    no = November
    de = December
    an = Any month

    Example: ja-mr is from January to March.

By-prod. (20 spaces). Artifacts and by-products eaten by animals.

    ca = Carrion
    eg = Eggs
    bo = Bones
    ga = Garbage

Cname. (40 spaces). Common name of animals or plants.

    Common name for a species will vary region by region. Some species that are not familiar to the public will not have common names, and in this case the data cell will be left blank. Forty spaces are reserved for a common name.

Fidelity. (10 spaces). Validation for an animal in a type in a state.

    1 = It occurs in type, but specifics are unknown
    h = Habitat (where a knowledgeable person would expect to see an animal)
    f = Fringe
        s = Summer
        w = Winter
        y = Yearlong
        m = Local movements

(continued on next page)

TABLE 5.2 (continued)

Directory for Tables (Relations) and Columns (Attributes) in Figure 5.1 with Menus for Data Cells

Forest. (25 spaces). Names of national forests by state.

    Alabama: Conecuh, Talladega, Tuskegee, Bankhead

    Arkansas: Ouachita, Ozark-St. Francis

    Arizona: Apache-Sitgreaves, Coconino, Coronado, Kaibab, Prescott, Tonto

    California: Angeles, Cleveland, Eldorado, Inyo, Klamath, Lassen, Los Padres, Mendocino, Modoc, Plumas, San Bernardino, Sequoia, Shasta-Trinity, Sierra, Six Rivers, Stanislaus, Tahoe

    Colorado: Arapaho, Grand Mesa, Gunnison, Pike, Rio Grande, Roosevelt, Routt, San Isabel, San Juan, Uncompahgre, White River

    Florida: Apalachicola, Ocala, Osceola

    Georgia: Oconee, Chattahoochee

    Idaho: Panhandle, Boise, Caribou, Challis, Clearwater, Nez Perce, Payette, Salmon, Sawtooth, Targhee

    Illinois: Shawnee

    Indiana: Hoosier

    Kentucky: Daniel Boone

    Louisiana: Kisatchie

    Maine: White Mountains

    Michigan: Hiawatha, Huron-Mainistee, Ottawa

    Minnesota: Chippewa, Superior

    Mississippi: Bienville, Delta, Desoto, Holy Springs, Homochitto, Tombigee

    Missouri: Mark Twain

    Montana: Beaverhead-Deerlodge, Bitterroot, Custer, Flathead, Gallatin, Helena Kootenai, Lewis and Clark, Lolo

    Nebraska: Nebraska

    Nevada: Humboldt-Toiyabe

    New Hampshire: White Mountain

    New Mexico: Carson, Cibola, Santa Fe, Lincoln, Gila

    New York: Finger Lakes

    North Carolina: Croatan, Nantahala, Pisgah, Uwharrie

    Ohio: Wayne

    Oregon: Deschutes, Fremont, Malheur, Mount Hood, Ochoco, Rouge River, Siskiyou, Siuslaw, Umatilla, Umpqua, Wallowa-Whitman, Willamette, Winema

    Pennsylvania: Allegheny

    South Carolina: Francis Marion-Sumter

    South Dakota: Black Hills

    Tennessee: Cherokee

    Texas: Angelina, Davy Crockett, Sabine, Sam Houston

    Utah: Ashley, Dixie, Fishlake, Manti-LaSal, Uninta, Wasatch-Cache

    Vermont: Green Mountain

    Virginia: Jefferson, George Washington

    Washington: Colville, Gifford Pinchot, Mount Baker-Snoqualmie, Okanogan, Olympic, Wenatchee

    West Virginia: Monongahela

    Wisconsin: Chequamegon, Nicolet

    Wyoming: Bighorn, Bridger Teton, Medicine Bow, Shoshone

TABLE 5.2 (continued)
Directory for Tables (Relations) and Columns (Attributes) in Figure 5.1 with Menus for Data Cells

Fungi. (10 spaces). Fungi eaten by animals.

mi = Mistletoe

hf = Hypogeous fungi

sf = Shelf fungi

Geog-area. (40 spaces). Geographical area.

E = East

C = Central

W = West

Ground. (10 spaces). Ground/underground material.

a = Logs/stumps

b = Duff-humus

c = Boulders/crevices

d = Rock field/talus material

e = Elevated cliff/ledge

f = Surface debris

g = Underground material

Home. (5 spaces). Home range of the species.

Order of Magnitude (Om) with base 10

Om	Acres
0.0	1
0.5	3
1.0	10
1.5	32
2.0	100
2.5	316
3.0	1,000
3.5	3,162
4.0	10,000
4.5	31,623
5.0	100,000

Example: A species with Om 3.5 would have a home range between 3,162 acres and Om 4.0 of 10,000 acres (Figure 2.9).

Inform. (40 spaces). Additional information.

Cell reserved for future use.

Currently a "d" in a cell indicates that there are data in at least one of the columns.

(continued on next page)

TABLE 5.2 (continued)
Directory for Tables (Relations) and Columns (Attributes) in Figure 5.1 with Menus for Data Cells

Invert. (20 spaces). Invertebrates eaten by animals.

    in = Insects (all categories)

    cr = Crustaceans

    sn = Snails

    sp = Spiders

    wo = Worms

    un = Unknown

Lform. (5 spaces). Life form of animals and plants.

    The life form of animals has six designations. The list can be added to as necessity dictates. This life form is not the same as life form used to associate species with habitat. Invertebrates are listed, but there is no information included for this category. However, information can be added when it is appropriate. Only one life form can occur in a data cell.

Animals

    am = Amphibian

    bi = Bird

    fi = Fish

    in = Invertebrates

    ma = Mammal

    re = Reptile

Plants

    tr = Trees

    sh = Shrubs

    gr = Grass

    fo = Forbs

    ca = Cacti

Local-area. (40 spaces). Local areas that can be designated mountain ranges, specific places, and so on.

Major-types. (10 spaces). Major types that occur in each state.

Microsites. (10 spaces). Microsite factors that affect survival and reproduction.

    1 = Tree nest

    2 = Ground nest/bed

    3 = Tree cavity

    4 = Ground cavity

    5 = In/on water

    6 = Tree perch, roost

    7 = Platform

Opening. (10 spaces). Opening (<10% canopy).

    a = Grass/forb

    b = Shrub

    c = Wet meadow

TABLE 5.2 (continued)
Directory for Tables (Relations) and Columns (Attributes) in Figure 5.1
with Menus for Data Cells

Ov-story. (10 spaces). Tree stand canopy

    a = Upper canopy

    b = General canopy/crown

    c = Low canopy

        1 = Dead tree/snag

        2 = Trunk

        3 = Bark

        4 = Isolated wolf tree

        5 = Edge (forest, openings, stages)

Pcode. (10 spaces). Plant code.

    See codes listed in Plants table in FAAWN.

Rcode. (5 spaces). Regional code.

    PNW = Pacific Northwest states

    PSW = Pacific Southwest states

    NRM = Northern Rocky Mountains

    CRM = Central Rocky Mountains

    SRM = Southern Rocky Mountains

    NPS = Northern Plains states

    SPS = Southern Plains states

    NES = New England states

    NLS = Northern lake states

    MCS = Midcontinent states

    SAS = South Atlantic states

    SCS = South Central states

Rname. (30 spaces). Regional names.

    Western states

        Pacific Northwest states

        Pacific Southwest states

        Northern Rocky Mountains

        Central Rocky Mountains

        Southern Rocky Mountains

    Central states

        Northern Plains states

        Southern Plains states

    Eastern states

        New England states

        Northern lake states

        Midcontinent states

        South Atlantic states

        South Central states

(continued on next page)

TABLE 5.2 (continued)
**Directory for Tables (Relations) and Columns (Attributes) in Figure 5.1
with Menus for Data Cells**

Sname. (40 spaces). Scientific name of animals and plants.

> The most important column in the table is the scientific name of a species. Sname is stable because of the accepted taxonomic naming system. Sname for a species is entered once in the table, and there are no duplicates in the data model. Forty spaces are reserved for a scientific name.

States (20 spaces)	**Scode** (5 spaces)
Alabama	AL
Arkansas	AR
Arizona	AZ
California	CA
Colorado	CO
Connecticut	CT
Delaware	DE
Florida	FL
Georgia	GA
Iowa	IA
Idaho	ID
Illinois	IL
Indiana	IN
Kansas	KS
Kentucky	KY
Louisiana	LA
Massachusetts	MA
Maryland	MD
Maine	ME
Michigan	MI
Minnesota	MN
Missouri	MO
Mississippi	MS
Montana	MT
North Carolina	NC
North Dakota	ND
Nebraska	NE
New Hampshire	NH
New Jersey	NJ
New Mexico	NM
Nevada	NV
New York	NY
Ohio	OH
Oklahoma	OK
Oregon	OR

TABLE 5.2 (continued)
Directory for Tables (Relations) and Columns (Attributes) in Figure 5.1 with Menus for Data Cells

States (20 spaces)	**Scode** (5 spaces)
Pennsylvania	PA
Rhode Island	RI
South Carolina	SC
South Dakota	SD
Tennessee	TN
Texas	TX
Utah	UT
Virginia	VA
Vermont	VT
Washington	WA
Wisconsin	WI
West Virginia	WV
Wyoming	WY

SuRe. (5 spaces). Survival and reproduction factors.
Fo = Food
Cover = Cover

Tname. (35 spaces). Names of types.	**Tcode** (5 spaces)
White-red-jack pine	10
Spruce fir	11
Longleaf-slash pine	12
Loblolly-shortleaf pine	13
Oak-pine	14
Oak-hickory	15
Oak-gum-cypress	16
Elm-ash-cottonwood	17
Maple-beech-birch	18
Aspen-birch	19
Douglas-fir	20
Hemlock-Sitka spruce	21
Ponderosa pine	22
Western white pine	23
Lodgepole pine	24
Larch	25
Fir-spruce	26
Redwood	27
Pinyon-juniper	28
Western hardwoods	29

(continued on next page)

TABLE 5.2 (continued)
Directory for Tables (Relations) and Columns (Attributes) in Figure 5.1 with Menus for Data Cells

Tveg. (30 spaces). Terrestrial vegetation parts eaten by animals.

    ac = Acorns
    co = Cones
    tw = Twigs (woody stems)
    bu = Buds
    po = Pollen
    le = Leaves
    ro = Roots
    ne = Needles
    ba = Bark
    nu = Nuts
    se = Seeds
    sa = Sap
    be = Berries
    fl = Flowers
    sh = Shoots
    nr = Nectar
    mv = Miscellaneous vegetation
    ho = Honey
    gr = Grass
    fo = Forbs
    fr = Fruits

Un-story. (10 spaces). Understory shrubs, saplings, herbaceous vegetation.

    a = Shrubs/saplings
    b = Herbaceous (grass/forb)

Vert. (20 spaces). Vertebrates eaten by animals.

    am = Amphibians
    bi = Birds
    fi = Fish
    sm = Small mammals
    mm = Medium-to-large mammals
    re = Reptiles
    Use commas to separate codes. Example: fi, sm, bi

Volume. (20 spaces). Volume as defined by stand size, density, and layers.

sa = Sapling	(1- to 4-in. dbh)
yo = Young	(4- to 10-in. dbh)
in = Intermediate	(11- to 17-in. dbh)
ma = Mature	(>17-in. dbh)
ol = Old forest	(mature + defined conditions)

TABLE 5.2 (continued)
Directory for Tables (Relations) and Columns (Attributes) in Figure 5.1
with Menus for Data Cells

a = Single story
b = Two story
c = Multistory
    1 = Thin   (10–40%)
    2 = Thick   (41–70%)
    3 = Closed   (>70%)
    + = Increase above indicated value

Water. (10 spaces). Water in a forest environment.

a = Riparian
b = Seeps/springs
c = Rivers/streams
d = Ponds/lakes
e = Moist site
f = Swamp
g = Near/in/on
h = Aquatic vegetation

Zip-code. (20 spaces). A combination of animal code (Acode), life form, state, and forest type (Tcode).

Note: Names in this table conform to the design names for Tables and Columns in Figure 5.1.

some cases, only a single entry is made in a data cell in a column, but most columns have a menu with several choices that relate to the column name. For convenience, Table 5.2 is included as a Word document on the accompanying CD so that a printed desk copy can be used to interpret codes in the FAAWN data model or any model that might be developed for a local area. Codes for all the entries are compatible for developing a data form that will reduce the time required to record field data and permit efficient loading of data into the data model.

The current data model is a work in progress. The relations in Figure 5.1 have gone through several hundred revisions to arrive at a satisfactory data model that combines physical components to associate animal species with forest types and to use codes and words that adequately describe the associations. Many items in the relational scheme are generalizations that need to be refined, and some of the codes and descriptions may seem odd because they are not in our common language. A major advantage of a relational database is that codes can be changed at any time without losing the original data, and this is an important feature when developing data models for local areas.

5.3.1 ANIMALS

Animals is a *core table* containing wildlife species in the United States. Every table in the data model is ultimately connected to Animals through a series of linked tables that contain basic life history information. Animals has a *one-to-one relation* with Food and a *one-to-many link* with Life-Needs and States. Several one-to-one relationships can be combined into one table, but the splitting of one table into two helps maintain a separation of subject matter.

The width of a column has to be specified when a table is created. In addition to text, there are other format choices, such as computed, number, memo, currency, and so on. A default value is 255 spaces for text. When first developing a column, it is best to accept the default value. Later, it is more efficient to change width to a specific value depending on the potential length of a data string. Leaving a default value is wasted space when text string characters are low, such as in Animals. The entire table uses 135 spaces for the text as shown, but if the default were used, the software would reserve 2,040 spaces for seven columns. A good practice is to increment five spaces at a time to set the needed value. Additional space can always be added. In Animals, there are seven columns of information relating to an individual species. However, this table can be expanded by adding columns when other information is needed, such as legal status (endangered or threatened species), subspecies designations, or a common name alias.

Conforming to the six design criteria discussed will create trade-offs in how data are stored and retrieved from tables. Trade-offs can be made without deviating from restrictions, depending on how the data are displayed. For example, the following four cases illustrate different ways of storing or displaying information.

Case 5.1

Species	Vegetation Type
Mule deer	Ponderosa pine
Mule deer	Mixed conifer
Mule deer	Pinyon juniper
Mule deer	Spruce-fir

Case 5.2

Species	Vegetation Type
Mule deer	pp, mc, pj, sf

Case 5.3

Species	pp	mc	pj	sf
Mule deer	x	x	x	x

Case 5.4

Species	Vegetation Type
Mule deer	pp, mc,
	pj, sf

A practical problem is one of storing data in tables, columns, and rows so that ultimately it can be printed in a report on an 8.5-by-11-inch sheet of paper. To summarize large numbers of types, Case 5.4 can be created by using Case 5.2, increasing the row depth, and setting the row width narrower to force the vegetation types into a vertical order. Case 5.4 is a version of Case 5.2 where many abbreviated types can be entered into one data cell and aligned vertically instead of horizontally as in Case 5.2. The Species column will contain only one name but the vertical list can contain many abbreviated Vegetation Types depending on the column width.

Case 5.1 provides the information required as a single topic in a data cell, but mule deer has to be entered for each vegetation type in which it occurs. This table conforms exactly to relational theory. In Case 5.2, mule deer is entered only once in the Species column and all vegetation types (multiple topics) are entered in one column with a comma delimiter; it is efficient for four to six vegetation types or normal conditions. Case 5.3 is not efficient because it has a column for every vegetation type. This could be a problem if there is a large number of types.

Case 5.1 is an acceptable table, but the design can lead to data redundancy and to visual clutter if there is a long list of vegetation types for each animal. However, Case 5.1 is preferred because of its value in joining information from two tables containing connecting and identical columns. A mixture of Cases 5.1 and 5.2 will most likely be the best design for many data models. Most of the design and format for Animals is applicable to other tables in the data model and is an example of an efficient design and format that can be easily changed.

5.3.2 FOOD

Individual food items by different food groups are contained in a Food table. Because of the large number of individual items eaten by a given species, it is more convenient to group items by animal and plant life forms to reduce the complexity of the relation. The menu for Food items accounts for vertebrates, terrestrial vegetation, by-products of life forms, fungi, invertebrates, and aquatic vegetation. If available, users can add data to this table. The Food table contains information for 561 of the species accounted for in the Animals table.

5.3.3 LIFE-NEEDS

The Life-Needs relation brings together the association of an animal with and use of forest components. The structural components (Table 1.1) of a forest are the first level of abstraction when identifying food and cover areas. These layers are columns (attributes) with descriptors. Each data cell in the SuRe (survive and reproduce) column must have a food or a cover designation so that each species will have a food and cover field (row) identified. These two designations are identified in the Life-Needs table under the SuRe column as Fo (Food area) and Cover (Cover area). These two entries have different numbers of letters as a visual aid in identifying food and cover rows on a printed page.

A second level of attributes is contained in the menu for each column. In some cases, there is a third level that can be identified as a grouping. All of these attributes could be contained in other columns, but different levels in the current design can be separated by using the Access *filter* menu, without adding more columns. *Microsite* is an attribute containing descriptors for cover and shelter that affect survival and reproduction of a species. Common name (Cname) and Life Form (Lform) have been entered into the Animals core table, but it is convenient to continue to use Cname and Lform in other tables for ease in identification, although this can be done with the Acode link. While this violates the one-time entry protocol, convenience to be able to recognize a common name, without looking up the name by an Acode reference, overrides protocol.

Life-Needs is where environmental factors come together to indicate how much defined habitat overlap exists from one animal to another. While there is considerable information on some species, mostly those that have been traditionally studied as game animals, the information on nongame and threatened and endangered animals is still developing, so the data model has many vacant spaces in its data cells. Data for categorizing how wildlife is associated with forest components was patterned after a species habitat profile (SHP) system developed for a mainframe computer (Patton 1978) and the Life-Form system for the Blue Mountains of Oregon and Washington (Thomas 1979).

The terms *food* and *cover* are well established in the wildlife literature, and there is a general understanding that cover refers to security or protection. In many cases, cover is associated with vegetation. However, wildlife also use nonvegetation components such as rocks, downed woody material, caves, and water for protection. Cover can be separated into roosting, resting, thermal, breeding, and so on when information is available and there is a need for a finer resolution. More than one vertical forest layer can be associated with a single species. The horizontal components have descriptive characteristics across the forest floor that are associated with the vertical strata in horizontal space. Conveniently for managers and researchers, forest ecosystems with stands and open areas described by vertical and horizontal components define a matrix with columns and data cells in a table.

5.3.4 STATES

For practical reasons, some species were excluded from States based on the definition of forest animals. Choices had to be made to eliminate those that were not directly related to trees and forest conditions. Eliminated species include turtles, shorebirds, waterfowl, and animals with subspecies designations. From a total of 1,578 species in Animals, there are 1,101 included in States and Life-Needs.

The definition of *habitat* includes a place where an animal lives. A place provides survival components of food and cover as listed in the Life-Needs relation. In a forest, the place can be hierarchically identified as Region and State for a low level of resolution. In the States table, the potential occurrence of an animal within a state has been developed for over 13,000 combinations of individuals and states. This was done by overlaying the geographical range of an animal on state boundaries to develop an animal-state association. Information for the distribution of animals by geographical area, region, and state is more common and better documented than by animals in a forest type in a state. The States table is a place to start for developing individual checklists for each of the 48 continental states.

5.3.5 HYPOTHESIZED

To develop the Hypothesized relation, two tables (States and Types-States) were joined using state code (Scode) to link the columns. The joined table contains over 63,000 combinations in 10 columns of attributes for associating 1,101 animals in 48 states to 20 forest types. The animals listed for forest types can reasonably be expected to occur in the types in a given state. However, as discussed, animals in one type in one state may not occur in another state in the same type, particularly if the two states are a great distance apart. This problem can be narrowed somewhat by the grouping (regionalization) of states into 12 geographical regions (Table 5.2, Rcode). The regionalization provides a better resolution for associating animal species to types that cut across ecological barriers.

Validation has been done separately for animals in Types-States and in States (Figure 5.1), but not for all the animals that actually occur in a combined association. A remaining job is to validate each animal and its occurrence in a type in a state. The Fidelity column is where validation is indicated. If an animal occurs in a forest type in a state some time during a year, then a "1" for yes is entered in the column. In addition, a degree of association can be included to replace the "1" to indicate whether the type is habitat (h) or a fringe (f). A delimiting comma is used to separate the entries (e.g., h, y), thereby indicating habitat and yearlong occurrence. Ultimately, the 63,000 combinations can be reduced to just those species that actually occur in a type in a state. Local validation of occurrence is important to reduce generalizations from region to region for those species for which validation is missing. This can be done by the reader for a specific state or local area from the list of potential associations of animals in the Types-States table. The Local-Area column is for further dividing a state into identifiable geographical areas such as mountain ranges or watersheds.

One attribute in the Hypothesized relation is a column named Zip-code. Zip-code is the combined codes of the life form, state, and type of an animal into one formula that represents the address where an animal is found; it is the "place" component of habitat. A zip code provides a considerable amount of information in one animal-habitat formula and can be a useful tool for analysis and assessment of management areas. Because the zip code is a condensed code, it can also be used as a summary of information for reporting purposes.

5.3.6 REGIONS

The Regions relation is a linking table that connects National, Types-States, and Hypothesized by Scode. Regions contains all the types that occur in a state in one data cell by major types, or all types that occur, in another data cell in a column. This type of table provides a summary for the 48 states (Table 3.2).

5.3.7 TYPES-STATES

Types-States is a table that lists types in each state and links Regions to Types-Codes with Scode and Tcode.

5.3.8 TYPES-CODES

Type names are long and use a lot of space; therefore, the information is entered only once in the type name (Tname) column. All forest types have a numerical code (Tcode) associated with the type name. The type name follows the naming convention on the Atlas Map of 1967 and the Society of American Foresters (SAF) (Eyre 1960) forest cover type designation. However, the Tcode used in the data model does not follow the codes shown on the 1967 map. This is because of the need for format continuity in using a relational system. Codes beginning with a "0" or having only one digit such as a "1" can sometimes cause problems when using SQL; therefore, the Tcodes are all two-digit numbers from 10 to 29.

5.3.9 PLANTS-TYPES

Pcodes in Plants serve the same purpose as animal codes and follow the naming convention of the Acode. Like the animal Acode, the Pcode is derived from the scientific name of plants. Pcode links Plants to the types that are listed in the Plants-Types table.

5.3.10 PLANTS

The scientific and common names of trees that occur in the 20 forest types are shown in the Plants relation. Additional information can be added by creating new columns for life history, medicinal use, specific uses for habitat restoration, and so on. Plants other than trees can be listed in the table as well as those associated with other land cover (grassland, shrubland, etc.).

5.3.11 NATIONAL

There are 138 National forests in the 48 continental United States (Table 3.3). However, not all states have a national forest. Forests for each state are shown by name and state where they occur. Because of the links between tables, it is possible to generate wildlife checklists for each national forest after identifying forest types that occur in each national forest.

5.4 A WAY OF LEARNING

5.4.1 USING THE DATA MODEL

The enclosed CD contains a data model (FAAWN) relating forest attributes to the wildlife needs for 1,101 animals in 20 forest types and 48 states in the United States. The data model is not a turnkey application that is ready to use for all areas. To use the data model, it must be copied from the CD to a computer that has Microsoft Access 7 on a hard drive.

Readers are encouraged to invest in one of the many books that provide the basics for developing an Access database. However, a copy of a book about Access that might be found in a used bookstore will contain sufficient information to learn the fundamentals of relational database development and data manipulation. Numerous free tutorials for Access can be found online, but one worth considering is by the Goodwill Community Foundation, Inc. Sign-up is required to use a free tutorial at their Web site: http://www.GCFLearnFree.org.

5.4.2 EXAMPLES OF OUTPUT

The purpose of a database is to store information that can be retrieved to answer questions. Some of the types of information that wildlife biologists and foresters can retrieve from the current FAAWN data model include the following examples, which are commonly used for planning and reporting purposes:

- A list of wildlife species in the United States with their scientific name and life form. The species list can be grouped by life form.
- A list of wildlife species that potentially can be found in each of the 48 continental states. The state list of species can be grouped by life form in each state to show diversity.
- A list of the SAF forest types in the United States that can be found in each state.
- A list of tree species for each forest type in each state.
- A list of wildlife species for each state and the potential occurrence of species grouped by life form by forest type.
- A grouping of species by their need for different forest components for food and cover, such as species that use tree cavities for protection or tree crowns for food.
- Information is available in the 11 tables that can be linked in ways to explore the development of adaptability/vulnerability indexes and comparisons of species numbers and life forms for states, regions, and types for diversity indexes.
- Information is available to group species by different components that are used for food and cover.

These examples are not all that can be extracted from the data model, but they provide an indication of just how powerful a data model can become as a management tool for wildlife biologists and foresters. Some output examples from the database are provided on the enclosed disk.

5.4.3 SUGGESTED ADDITIONS

There are several topics that would be good additions to FAAWN to make the model more useful for management, for example:

- The addition of a Legal column in the Animals table to indicate the threatened, endangered, or other value of animals.
- A table for References that would link animal species to documents for validation of their food and cover needs.
- A table for Management Practices that would link animals to specific management practices for restoration or habitat maintenance.
- A table for Fire that would link the effects of fire, both positive and negative, to animal species.

There is information available in the published literature on the four topics listed. The documentation opportunity for including these for the Southwest has begun but sufficient progress has not been made for examples to be included in the current FAAWN data model.

5.4.4 ENTERING AND EDITING DATA

Entering and editing data are continuous to improve the data model design and quality of information. Many errors occur when information is first entered into a data cell, and these errors may not show up for a long time. One of the most common mistakes is entering the wrong code from a menu or the right code in the wrong data cell. These types of errors are often difficult to detect, but sometimes they become obvious when looking at patterns on the screen. If something seems out of place, then it probably is and needs to be checked for accuracy.

The FAAWN data model contains sufficient information to explore by asking questions, linking tables in a query, and filtering from a course to a finer level of resolution to get specific information. For more extensive answers to nested questions, the reader is encouraged to query information for his or her state and compare the output to that of other states. All of these activities are a part of the learning experience that will help to understand just how useful a data model could be if all the tables and columns were completely filled.

A good practice in making detailed changes in a table in the data model is to copy and paste the table for a second version. Rename the second copy and put a 99 in front of the name to show that it is an experimental copy of the original. Experimental tables provide a way of checking a command, such as when combining tables in a query, using the find-and-replace option of changing codes, or just to see what the results are for a query before you initiate the command in the original table. It is also a good practice to have an experimental copy of the original data model to try different commands, especially when changing relationships of linking tables.

5.5 USING THE COMPUTER DISK

Insert the enclosed CD in an appropriate drive on your computer. There are five files on the disk:

1. **FAAWN in Microsoft Access.** Copy this to the hard drive on your computer.
2. **Table 1.2.** This is the attributes menu for components in the Life-Needs table. Print a copy from the hard drive for a desk reference. You will need this information before opening the Life-Needs table.
3. **Table 5.2.** This is the directory for tables and columns in Figure 5.1. Print a copy from the hard drive for a desk reference.
4. **Figure 5.1.** This is the FAAWN relational data model design with table names and column headings. Print a copy from the disk for a desk reference.
5. **Getting started.** Basic instructions are provided for using the FAAWN data model in Microsoft Access. Print a copy for a desk reference.

Note: Because of the large database file, your computer may be slow in copying files to the hard drive.

KNOWLEDGE ENRICHMENT READING

Codd, E. F. 1990. *The relational model for database management: Version 2.* Addison-Wesley, New York.

Groh, R. R., J. C. Stockman, G. Powell, et al. 2007. *Microsoft Access 2007.* Wiley, Indianapolis, IN.

MacDonald, M. 2006. *Access 2007. The missing manual.* Pogue Press, O'Reilly Media, Sebastopol, CA.

Stockman, J. C., and A. Simpson. 2007. *Microsoft Office Access 2007 VBA programming for dummies.* Wiley, Hoboken, NJ.

REFERENCES FOR VALIDATING FAAWN INFORMATION

Behler, J. L. 1979. *Field guide to North American reptiles and amphibians.* Alfred A. Knopf, New York.

Brown, E. R. (Tech. Ed.). 1985. *Management of wildlife and fish habitats in forests of western Oregon and Washington.* U.S. Forest Service R6F&WL 192, Parts 1 and 2. PNW Region, Portland, OR.

Bull, J. 1977. *Field guide to North American birds: Eastern region.* Alfred A. Knopf, New York.

Clark, W. S. 1987. *Hawks.* Houghton Mifflin, New York.

Conant, R. 1998. *Reptiles and amphibians: Eastern/central North America.* Houghton Mifflin. New York.

DeGraff, R. M. (Tech. Coord.). 1978. *Management of southern forests for nongame birds: Proceedings of the workshop.* General Technical Report SE-14. U.S. Forest Service, Asheville, NC.

DeGraff, R. M., and K. E. Evans. 1979. *Management of north central and northeastern forests for nongame birds: Proceedings of the workshop.* General Technical Report NC-51. U.S. Forest Service, Minneapolis, MN.

DeGraff, R. M., V. E. Scott, R. H. Hamre, et al. 1991. *Forest and rangeland birds of the United States: Natural history and habitat use.* Agriculture Handbook No. 688. U.S. Forest Service, Washington, DC.

DeGraff, R. M., and N. G. Tilghman. 1980. *Management of western forests and grasslands for nongame birds: Proceedings of the workshop.* U.S. Forest Service General Technical Report INT-86. Intermountain Region, Ogden, UT.

Ehrlich, P. R., D. S. Dobkin, and D. W. Naeem. 1988. *The birder's handbook.* Simon and Schuster, New York.

Eyre, F. H. (Ed.). 1980. *Forest cover types of the United States and Canada.* Society of American Foresters, Washington, DC.

Feldhamer, G. A., B. C. Thompson, and J. A. Chapman. 2003. *Wild mammals of North America: Biology, managements and conservation.* Johns Hopkins University Press, Baltimore, MD.

Fleming, C., and B. von Halle. 1989. *Handbook of relational database design.* Addison-Wesley, New York.

Garrison, G. A., A. J. Bjustad, D. A. Duncan, et al. 1977. *Vegetation and environmental features of forest and range ecosystems.* Agriculture Handbook 475. U.S. Forest Service, Washington, DC.

Hall, E. R. 1981. *The mammals of North America*, Volumes 1 and 2. John Wiley and Sons, New York. Second edition. Reprint by Blackburn Press, Caldwell, NJ.

Hoover, R. L., and D. L. Wills (Eds.). 1984. *Managing forested lands for wildlife.* U.S. Forest Service, Rocky Mountain Region and Colorado Division of Wildlife, Denver, CO.

Kircher, J. C. 1993. *Ecology of western forests.* Houghton Mifflin, New York.

Kircher, J. 1998. *Rocky Mountain and southwest forests.* Houghton Mifflin, New York.

Kircher, J. C. 1998. *Eastern forests.* Houghton Mifflin, New York.

Kircher, J. C. 1998. *California and Pacific Northwest forests.* Houghton Mifflin, New York.

Kroenke, D. M., and D. E. Nilson. 1986. *Database processing for microcomputers.* Science Research Associates, Chicago, IL.

National Geographic Society. 1983. *Birds of North America.* National Geographic Society, Washington, DC.

Patton, D. R. 1975. *Computer print outs. Run Wild System 2000 storage and retrieval system.* U.S. Forest Service Rocky Mt. Forest and Range Experimental Station, Wildlife Research Unit RM-1710, Arizona State University Campus, Tempe.

Patton, D. R. 1978. *Run Wild, a storage and retrieval system for wildlife habitat information.* General Technical Report RM-51. U.S. Forest Service, Fort Collins, CO.

Peterson, R. T. 1980. *A field guide to eastern birds.* Houghton Mifflin, Boston.

Peterson, R. T. 1990. *A field guide western birds.* Houghton Mifflin, Boston.

Petrides, G. A. 1992. *A field guide to western trees.* Houghton Mifflin, Boston.

Petrides, G. A. 1988. *A field guide to eastern trees.* Houghton Mifflin, Boston.

Reid, F. A. 2006. *Mammals of North America.* Houghton Mifflin, Boston.

Scott, V. E., K. E. Evans, D. R. Patton, et al. 1977. *Cavity nesting birds of North American forests.* Agriculture Handbook 511. U.S. Forest Service, Washington, DC.

Smith, D. R. (Tech. Coord.). 1975. *Proceedings of symposium on management of forest and range habitats for nongame birds.* General Technical Report WO-1. U.S. Forest Service, Washington, DC.

Stebbins, R. C. 2003. *Western reptiles and amphibians.* Houghton Mifflin, Boston.

Sutton, A., and M. Sutton. 1985. *Eastern forests.* Alfred A. Knopf, New York.

Szaro, R. C., K. S. Severson, and D. R. Patton (Tech. Coords.). 1988. *Management of amphibians, reptiles, and small mammals in North America.* General Technical Report RM-166. U.S. Forest Service, Ft. Collins, CO.

Thomas, J. W. (Ed.). 1979. *Wildlife habitats in managed forests, the Blue Mountains of Oregon and Washington.* Agriculture Handbook 553. U.S. Forest Service, Washington, DC.

Udvardy, M. D. F. 1977. *Field guide to North American birds, western region.* Alfred A. Knopf, New York.

Verner, J., and A. S. Boss (Tech. Coord.). 1980. *California wildlife and their habitats, Western Sierra Nevada*. General Technical Report PSW-37. U.S. Forest Service, Berkeley, CA.

Whitaker, J. O. 1980. *Field guide to North American mammals*. Alfred A. Knopf, New York.

Whitaker, J. O., Jr., and W. Hamilton Jr. 1998. *Mammals of the eastern United States*. Comstock, Cornell University Press, Ithaca, NY.

Whitney, S. 1985. *Western forests*. Alfred A. Knopf, New York.

Zabel, C. J., and R. G. Anthony (Eds.). 2003. *Mammal community dynamics, management and conservation in the coniferous forests of western North America*. Cambridge University Press, Cambridge, UK.

REFERENCES

Codd, E. F. 1970. A relational model of data for large shared databanks. *Commun. ACM* 13(6): 377–387.

Fleming, C., and B. von Halle. 1989. *Handbook of relational database design*. Addison-Wesley, New York.

Kroenke, D. M., and D. E. Nilson. 1986. *Database processing for microcomputers*. Science Research Associates, Chicago, IL.

Patton, D. R. 1978. *Run Wild, a storage and retrieval system for wildlife habitat information*. General Technical Report RM-51. U.S. Forest Service, Fort Collins, CO.

U.S. Congress. 1974. Forest and Rangeland Renewable Resources Planning Act. Public Law 93-378. U.S. Government Printing Office, Washington, DC.

Appendix: Conversion Factors

Multiply	By	To Convert to
Length		
chains (ch)	20.1168	meters
feet (ft)	0.30480	meters
inches (in.)	2.5400	centimeters (cm)
kilometers (km)	0.62137	miles
meters (m)	3.28083	feet
meters (m)	0.04971	chains
millimeters (mm)	0.03937	inches
miles (mi)	1.60934	kilometers
yards (yd)	0.91440	meters
Area		
acres (ac)	0.40468	hectares
acres (ac)	43,560	square feet
hectares (ha)	2.47104	acres
hectares (ha)	10,000	square meters
square feet (ft^2)	0.09290	square meters
square kilometers (km^2)	0.38610	square miles
square kilometers (km^2)	100	hectares
square kilometers (km^2)	247.104	acres
square meters (m^2)	10.76387	square feet
square miles (mi^2)	2.58999	square kilometers
square miles (mi^2)	640	acres
Basal Area		
square feet/acre (ft^2/ac)	0.2296	square meters/hectare
square meters/hectare (m^2/ha)	4.3560	square feet/acre
Weight		
kilograms (kg)	2.20462	pounds
kilograms/hectare (kg/ha)	0.8922	pounds/acre
pounds (lb)	453.592	grams (g)
pounds/acre (lb/ac)	1.1208	kilograms/hectare
Volume		
board feet (bd ft)	0.00566	cubic meters
board feet/acre (bd ft/ac)	0.01399	cubic meters/hectare
cubic feet (ft^3)	0.0283	cubic meters
cubic feet/acre (ft^3/ac)	0.06997	cubic meters/hectare
cubic meters (m^3)	35.3145	cubic feet

(continued on next page)

Multiply	By	To Convert to
cubic meters (m³)	176.57	board feet
cubic meters/hectare (m³/ha)	71.4571	board feet/acre
cubic meters/hectare (m³/ha)	14.2913	cubic feet/acre
gallons (gal)	3.785	liters
liters (l)	0.26417	gallons

	Temperature	
Centigrade	$0.5556 (F - 32)$	Fahrenheit
Fahrenheit	$32 + (1.80 C)$	Centigrade

Index

A

Abiotic components, 18
Abrupt edges, 24
Abundance, 24
    animals, based on frequency of occurrence, 218
Acode link, 246
Adaptive resource management, 124, 195–196
Advanced very-high-resolution radiometer (AVHRR), 27
Aerial photography, 142, 203
Age-specific life table, 82
Agents of change, 38
Allogenic succession, 38
Allotment analysis reports, 203
Alpha diversity, 17
Alpha species, 92
Amphibians, 5
    effects of fire on, 56
Animal counting, 215–216
Animal density
    and carrying capacity, 81
    crude *vs.* ecological, 70
    seasonal fluctuations and hunting, 133
    and soil fertility, 35
    and territories, 68
Animal name code, 234
Animal numbers, estimating, 215
Animal signs, 216–218
Animal species
    dependence on plants, 2
    inability to produce own food, 33
    as integral parts of ecosystem, 3
    variance in ecological amplitudes, 2
Animal succession, 37
Animal unit equivalents, 143–144
Animals
    air temperature and death, 55
    fire impacts, 55
    vehicle collisions with, 57
Animals core table, in FAAWN, 243–245
Antelope
    effects of fences on, 58
    water requirements, 89
Anthropomorphism, 7
Apache National Forest, xiii
Aquatic coordination and restoration measures, 158–159
Aquatic vegetation, eaten by animals, 234
Arctic warming, 46
Area conversion factors, 255
Area regulation, 121
Artifacts, 34
    eaten by animals, 235
Aspen-birch forests, 103, 110
    common wildlife species, 110
Attributes, 229
    FAAWN directory, 234–243
    in relationship systems, 229
Autecology, 10, 39
Autogenic succession, 38
Autotrophism, in plants, 33
Availability, of food resources, 86
Axioms, 1–3, 24, 33, 35, 42, 134

B

Background information checklist, 203
Bacterial diseases, 59
Barbed wire fencing, 58, 125
Basal area (BA), 185
    conversion factors, 255
    converting tree diameter into, 219
Bayes's theorem, 183
Beaver Creek Watershed, xiii
Behavior, 68
Beta species, 92
Bighorn sheep, water requirements, 89
Biogeochemical cycles, 32
Biogeography, 164, 165
Biological diversity, 17, 20–23, 21, 174
Biological legacies, 45
Biological microsites, 20
Biology, 66
    and behavior, 68
    carrying capacity, 81–82
    demographic vigor, 85
    and ecological niche, 69
    fluctuations, 83–85
    inhibited growth model, 78–81
    life table, 82–83
    and population characteristics, 69–72
    rates of population change, 74–78
    of reproduction, 66–68
    and territory, 68–69
    uninhibited growth model, 73–74
    in wildlife management, 128–134

Biomes, 25
Biotic components, 18, 31
Biotic provinces, 25
Biotic pyramids, 31
Birds, 5
Birth rates, 70
Birth stage, 11
Bison populations, human effects on, 64
Bobwhite quail
    protein requirements, 87
    resting cover for, 88
Body size, reproductive characteristics associated
    with, 93
Botfly larvae, 126
Botulism, 60
Bounty hunters, 128
Branching chains, 6
Breeding cover, 246
Breeding season, 66, 235
    uninhibited growth model, 73
Broadleaf trees, 10
Browse species, 190
    classification by form and age class, 193
    fire management and, 125
Bubonic plague, 59–60
Bureau of Indian Affairs (BIA), 47
Bureau of Land Management, 143

C

Canopies, 15
Canopy closure, 185
Carbon cycle, 32
Carrying capacity, 81–82, 143
Cause-effect relationships, 173, 191
Cavity-nesting birds, 148, 149
    snag management for, 170
Center of activity, 92
Central states, forest cover types, 102
Centrum, 6–7, 124, 197
    humans as hunters in, 64
Change agents, 38
Changes
    in bird species, 43
    in mammal species, 43
    over time, 42
Choice, 23
Circular plot sizes, 211
Classification
    browse plant species, 193
    forest wildlife, 4
    vegetation, 25–27
Clean agriculture, increases in, 65
Clearcutting, 116, 117, 118–119, 122
    to achieve natural landscape pattern with
        edge, 118
    in Pacific Northwest, 118

Climate, 45
Climate change, 45–47
Climate change Science Program (CCSP), 46
Climax patterns, 39, 134
    stability of, 39
Climax stage, 37, 38
Cloud seeding, 124
Coarse woody debris, 29–30
Codominance, 25, 26
Cohorts, 73
Cold water, 13
Commercial forests, 101
Commercial thinning, 121
Common knowledge axioms, 1–3
Common names, 246
Communities, 30
Competition
    animal unit equivalents, 144
    food habits overlap, 144
    integrated management considerations,
        143–145
Competitive exclusion, 69
Complementarity principle, 48
Complexity, and stability, 16–17, 35
Composite life table, 83
Conceptual models, 180
Conflicts
    between human and wildlife habitats, 196
    in integrated management, 145
    over clearcutting, 116
    over human behavior, 127
    over livestock grazing, 115
    over old growth habitats, 121
    over tree harvesting, 115
Coniferous forests, 27, 101
Conservation organizations, 127
Constancy, 35
Consumer biomass, 31
Continental ecosystems, 164
Continuous forest inventory (CFI), 48
Continuous grazing, 143
Controlled grazing, 135
Conversion factors, 255–256
Cool water, 13
Cottontail rabbit, low-mobility species, 67
Cover, 88–89. *See also* Food and cover
    defining minimum forest blocks, 165
    nonvegetation used as, 246
    for white-tailed deer, 88
    wildlife loss of, 121
Cover map, 214
Cover type, 27, 88
Coyotes, seasonal food habits, 86
Critical habitat, 9
Crown area, 219
Crude birth rate, 70
Crude density, 70

Cruising radius, 92
Cutting cycle, 123–124
Cyclic fluctuations, 84

D

Darwin, Charles, 164
Data cell menus, for FAAWN data model,
    234–243
Data cells, 230
Data entry/editing
    in FAAWN, 250
    and quality control, 228
Data linking, 228
Data manipulation, manual *vs.* computer, 168
Data models, 5, 227, 229
    validation and testing, 233
Data types, 230
Database design
    evaluating and testing, 232–233
    links rules, 231–232
    tradeoffs, 232
Database management systems (DBMS), 194
Database schemes, 227
Databases, 187, 188
DDT, 61
Dead trees, 147
    food and cover uses, 149
Death, 11
Decadent stands, 42
Deciduous trees, 10, 27
Decimating factors, 6
Decision making, 230
    decision trees and, 181
    expert systems and, 181–182
    in habitat models, 180–181
Decision support systems, 203–204
Decision trees, 181
Decomposers, 32
Decreasers, 191, 192
Deer populations
    in featured species model, 171
    pellet count technique for estimating, 217
    and predator-prey relationships, 63
Deferred grazing, 143
Deforestation, 32
    conversion to cropland and urban uses, 65
Demographic vigor, 85
Density, 24
    of water, 12
Density-dependent factors, 80
Density-independent factors, 80
Deterministic models, 180
Directly acting components, 6, 7
Disease-producing agents, 59–61
Disease vectors, 60

Diseases
    difficulty of treating wild animals for, 126, 127
    disease-producing agents, 59–61
    herbicides and, 61
    insecticides and, 61
    pesticides and, 61
    as wildlife hazard, 59
    in wildlife management, 125–126
Diversity, xiv
    biological, 20–23
    component arrangement, 20
    components, 18–20
    defined, 16
    ecosystem, 22
    edge and, 23–24
    forest components attributes menu, 21
    in forest environments, 16–17
    genetic, 22
    horizontal and vertical, 17
    landscape, 22
    layers, 17–18
    restoring and maintaining, 146–159
    species, 22
    structural components in forests, 19
    visual representation, 19
Dominance, 25, 26
Dominant forest types, 105–106
Double clutching, 130
Douglas-fir forests, 110–111
    common wildlife species, 111
    in Western states, 104
Drought, 58
Dwarf shrubland, 25
Dying trees, 148
Dynamic equilibrium, 82

E

Early successional species, 4
Earth ecosystem, 164
Eastern states, forest cover types, 102, 103–110
Ecological amplitude, 2, 36–37, 42
Ecological axioms, 1–3
Ecological carrying capacity, 82
Ecological density, 70
Ecological diversity, 16
Ecological dominants, 3, 40
    humans as, 127
Ecological indicators, 172, 173
Ecological longevity, 67
Ecological management factors, 172–173
    biological diversity, 174
    edge, 174–175
    gap analysis, 177
    habitat corridors, 176–177
    indicators, 173–174
Ecological niche, 69

Ecological time, 36, 44, 164
  cover changes in, 89
Economic carrying capacity, 82
ECOSIM model, 170
Ecosystem diversity, 22
Ecosystem functioning, 188
Ecosystem management strategies, 163
  ecological factors, 172–177
  habitat evaluation and monitoring, 188–194
  habitat models, 177–188
  integrated management, 194–196
  inventory, 205–209
  planning, 196–205
  sampling, 209–219
  wildlife habitat relationships, 166–172
Ecosystems, 163–164
  habitat area, 164–166
  landscapes in, 164
  local and formational, 164
Edge effect, 23
  obtaining through clearcutting, 118
Edge habitats, 4
Edges, 23–24, 174–175
  comparison of patterns, 175
  creation by riparian vegetation, 15
Elk
  adaptation to roads, 57, 58
  effects of fences on, 58
  habitat management guidelines, 166
  pellet count technique for estimating, 217
  and riparian meadow habitats, 58
  thermal cover for, 88
Elm-ash-cottonwood forests, 103, 109–110
  common wildlife species, 110
Emigration, among wildlife, 70
Empirical models, 180
Endangered habitats, 16
Endangered species, 4, 22, 173
  advantages of small management units for, 165
  controversies, 115
  effects of severe weather on, 58–59
Endangered Species Act, 198
Energy flow, 34
  in ecosystems, 33–34
Energy pathways, 34–35
Engineering coordination and restoration measures, 156–157
Engines of ecosystems, 34
Environmental factors, 55
Epilimnion, 13
Erosion, due to recreational campground use, 136
Escape cover, for squirrels, 88
Estimated protection levels, 71
Eutrophication, 13–14

Even-aged stands, 116, 117
  creating through clearcutting, 116
  regenerating through seed tree harvesting, 120
  regulation to produce, 122
Evergreen trees, 10
Evolutionary viability, 72
Exclosure location, 203
Exotic species, 4
  introducing through stocking, 129
  as pests, 129
Expert systems, 181–182
Exponential growth, 73
Extinction rates, 22
Extreme events, 47

F

FAAWN. *See* Forest Attributes and Wildlife
      Needs (FAAWN) data model
Fall turnover, 13
Family resilience, 72
Farm species, 4
Featured species management, 170–171
Fecundity, 66, 67
  mobility and body size, 93
  and natality, 70
Fences
  as man-made hazards, 125
  as wildlife hazards, 58
Feral species, 4
Fidelity, 235
Field data records (FDR), 206
Fir-spruce forests, 104, 112–113
  common wildlife species, 113
Fire-adapted species, 57
Fire-created habitats, 125
Fire-dependent species, 57
Fire hazards, 55–57
  dead trees and, 28
Fire-impervious species, 56, 57
Fire-intolerant species, 56, 57
Fire management techniques, 125
First law of thermodynamics, 34
Fish, 5
  effects of fire on, 56
  sensitivity to temperature, 13
Fish and Wildlife Coordination Act, 197
Fish and Wildlife Service, 189
Flat fluctuations, 84
Fluctuations, 83–85
  cyclic, 84
  flat, 84
  irruptive, 84
Food and cover, 246
  dead trees and snags as, 149
  forbs used for, 151

grass and grasslike plants for, 152
North American trees used by wildlife for,
    147
shrubs as, 150
Food chains, 34
    indirect effect of pesticides in, 61
    role of plants, 34
    tassel-eared squirrel, 32–33
Food map, 214
Food palatability, 86
Food preference, 87
Food resources, 86–87
    animal protein requirements, 87
    available to predators, 64
    coyote seasonal food habits, 86
Food storage, 149
Food table, in FAAWN, 245
Food-to-cover ratio, 137
Food webs, 34
Foot-and-mouth disease, 59, 60
Forbs, 10, 150
    food and cover uses, 151
Forest and Rangeland Renewable Resources
        Planning Act, 9, 199, 233
Forest Attributes and Wildlife Needs (FAAWN)
        data model, xv, 114, 225–227
    Ac-Sc, 234
    Acode, 234
    Acres, 234
    All-types, 234
    Animals core table, 243–245
    Aq-veg, 234
    Born column, 234
    Breeds, 235
    By-prod, 235
    CD use, 250–251
    Cname, 235
    data cell menus, 234–243
    data entry and editing, 250
    data trade-offs, 244
    directory for relations and attributes, 234–243
    as educational tool, 225
    Fidelity, 235
    Food table, 245
    Forest column, 236
    Fungi, 237
    fuzzy codes in, 227
    Geog-area, 237
    good and not-so-good codes, 227
    Ground material, 237
    Home table, 237
    Hypothesized relation, 247
    Inform column, 237
    Invert column, 238
    learning through using, 248–250
    Lform column, 238
    Life-Needs relation, 245–246

Local-area column, 238
Major-types column, 238
Microsites column, 238
National table, 248
Opening column, 238
output examples, 249
Ov-story column, 239
pattern recognition, 227
Pcode column, 239, 248
Plants table, 248
Plants-Types table, 248
Rcode column, 239
Regions linking table, 247
relational data model development, 233–248
relational database systems, 228–233
relations in, 233
Rname column, 239
Sname column, 240
species exclusions, 246
States table, 240–241, 246–247
suggested additions, 249–250
SuRe column, 241
table and column directory, 234–243
table names and column headings, 226
Tname column, 241
Types-Codes table, 248
Types-States table, 248
Un-story column, 242
using, 248–249
Volume column, 242
Water column, 243
Zip-code column, 243
Forest components
    attributes menu, 21
    relations with animals, 225
    space, 91
Forest diversity, 16–17
    structural components, 19
Forest ecological systems, 30
    carbon cycle in, 32
    ecological amplitude, 36–37
    energy flow in, 33–34
    hierarchical organization, 30–31
    limiting factors in, 37
    nutrient and energy pathways, 34–35
    nutrient cycles, 32–33
    soil in, 35
    structure and processes, 31
    time and space in, 36
Forest ecosystems, 25
Forest edge, 18. *See also* Edges
Forest fragmentation, 164
Forest health, 44
Forest management, 114–115. *See also* Integrated
        forestry and wildlife management
    clearcutting, 116, 118–119
    cutting cycle, 123–124

even- and uneven-aged stands, 116
forest regulation, 121–122
integrating with wildlife management, 101
old forest conditions, 120–121
rotation, 122–123
seed tree harvesting, 120
selection harvesting, 120
shelterwood harvesting, 119
silvicultural systems, 115
thinning, 121
tree spacing, 121
tree species tolerance, 115–116
Forest plants, 10–11
life form, 10–11
Forest regeneration methods, 117
following harvesting, 115
Forest regulation, 121–122
to produce even-aged stands, 122
to produce uneven-aged stands, 123
Forest species, 4
Forest succession, 37–38
agents of change, 38
changes over time, 42–44
old forest conditions, 41–42
primary succession, 38–39
secondary succession, 39
tree structural stages, 39–40
Forest types, 26–27, 101–103, 226
aspen-birch, 103, 110
in Central states, 102
dominant, 105–106
Douglas-fir, 104, 110–111
in Eastern states, 102, 103–110
elm-ash-cottonwood, 103, 109–110
fir-spruce, 112–113
hemlock-Sitka spruce, 111
hierarchy example, 101–102
larch, 104, 112
loblolly-shortleaf-pine, 103, 108
lodgepole pine, 104, 112
longleaf-slash pine, 103, 106, 108
maple-beech-birch, 103
naming in FAAWN, 241
oak-gum-cypress, 103, 109
oak-hickory, 103, 109
oak-pine, 103, 108
pinyon-juniper, 113
ponderosa pine, 104, 111–112
redwood, 113
spruce-fir, 103, 106
by states, 105, 106
western hardwoods, 114
in Western states, 102, 110–114
western white pine, 104, 112
white-red-jack pine, 103
Forest wildlife species, 3
Forestland, 25, 26

Forests
as classrooms, xiii
as ecological systems, 30–37
international perspectives, xiv
Formational ecosystems, 164
Formations, 26
FORPLAN software, 203, 204
Frequency, 25
Frequency classes, 218
Frequency of occurrence, abundance of animals
based on, 218
Freshwater habitats, 12–13
Fuel breaks, 124
Fungal diseases, 59
Fungi-rodent relationships, 33
Fungi table, 237
Furbearers, 4
Fuzzy codes, 227

G

Galapagos Islands, 164
Gambel's quail, precipitation and reproduction, 90
Game, 4
Game and fish agencies, 135
Game animals
harvesting, 130
importance of shrubs, 150
nutritional requirements, 87
Game farm animals, 130
Game kill records, 204
Gamma diversity, 17, 22
Gamma species, 92, 165
Gap analysis, 177
Gap phase replacement process, 39
Genetic diversity, 22
Geographic data, 203
Geographic information systems, 194, 209
Geographical areas, 237
forest types, 101
Geographical range, 91
Geological time, 36
Gila National Forest, xiii
Global warming, 32, 47
Good forest management, 48
Governing policies, 199–201
Graminae family, 10
food and cover uses, 152
Grasses, 11
food and cover uses, 152
Grasslike plants, 11
Gray squirrel, scorecard for evaluating habitat, 192
Grazing
controlled, 135
positive effect on brushland-dependent
species, 145

Grazing allotments, 143
Grazing capacity, 81
Grazing permits, 145
Grazing rates, 145
Grazing systems, 143
Greenhouse effect, 32, 46
Ground material, 237
Ground-nesting birds, grazing effects on, 145
Group selection, 120
Grouse, nesting cover for, 88
Growth stages, 11
Guild matrix, pine-oak woodland birds, 172
Guilds, 170, 171

H

Habitat, 7–10
    definitions for FAAWN, 246
    by forest type, 114
    freshwater, 12–13
    loss of, 57
Habitat area, 164–166
Habitat capability models (HCMs), 169
    tassel-eared squirrel, 169
Habitat change monitoring, 194
Habitat characteristics, used by wildlife, 230
Habitat condition and trend, 190–193
Habitat corridors, 176–177
Habitat data, uses, 214–215
Habitat destruction, by humans, 64–65
Habitat dispersion, 138
Habitat evaluation and monitoring, 188
    browse plants classification, 193
    condition and trend, 190–193
    ecosystem functioning, 188
    habitat change monitoring, 194
    habitat quality and populations, 189
    habitat suitability index, 189–190
Habitat evaluation procedure (HEP), 189
Habitat generalists, 3, 92
Habitat indices, 185–186
Habitat management, 146
    guidelines for elk, 166
    for specific species, 134
Habitat manipulation, 135
Habitat models, 177–178
    basal area per acre, 185
    conceptual, 180
    decision making in, 180–182
    degree of detail, 179
    deterministic, 180
    dollar values required for particular species, 183
    empirical, 180
    factors to consider, 178–180
    habitat indices, 185–186
    inputs, manipulative functions, outputs, 179

pattern recognition, 183–184
    quantitative, 184
    as real-world representations, 178
    stand structure model, 186
    turkey habitat attributes sample data, 184
    wildlife data models, 186–188
    wildlife ranking by habitat quality, 182
Habitat populations, 189
Habitat quality, 182, 189
    wildlife species ranking by, 182
Habitat ratings, 182
Habitat scorecard, 192, 193
Habitat selection, 9
Habitat specialists, 3, 92
Habitat suitability, 9
Habitat suitability index, 169, 189–190
Habitat use, 9
Harvesting, of game animals, 130
Hazards, 55
    fences, 58
    fire, 55–57
    roads, 57–58
    weather, 58–59
    in wildlife management, 124–125
Healthy forest conditions, 44
    climate change, 45–47
    long-term productivity, 44
    natural events, 44–45
    sustainability, 45
Hedging, 193
Hemlock-Sitka spruce forests, in Western states, 104, 111
    common wildlife species, 111
Herbaceous plants, 11, 25
    wildlife use, 150
Herbage production, 190
Herbicides, 61, 126
Hiding cover, 30
Hierarchical dominance, 69
Hierarchy, in ecological systems, 30–31
High-intensity inventories, 207
High-mobility species, 67
High-value food, 87
Historical data, 203
Holistic approach, 39, 163
Hollow trees, 29, 147
Home range, 93, 237
    mammals, 92
    Om comparison to size of area, 94
    overlapping, 95
Horizontal diversity, 17
    in forest ecosystem, 18
Host-parasite relationships, 60
Human-wildlife conflicts, 64
Humans
    ability to modify succession, 40
    as ecological dominants, 3, 40, 127

effect on climate change, 32
and habitat destruction, 64–65
here-and-now time scale, 137
illegal hunting by, 65–66
impact on survival and reproduction, 64–66
predisposition to hunt, 131
role in animal diseases, 60
wildlife management factors, 127
Hunting
controversies, 131
human predisposition to, 131
population percentages removable by, 131
rationales, 133
regulation objectives, 130
Hunting perches, 149
Hunting preserves, 130
Hurricane Hugo, 45
Hurricanes, 58
Hydrology changes, 47
Hypolimnion, 13
Hypothesized table, 225, 226
in FAAWN, 247

I

Illegal hunting, 65–66
Immigration, among wildlife, 70
In the wild, 3
Increasers, 191, 192
Indicators, 173–174
Indirectly acting components, 6, 7
Individual survival, 72
Individuals, 30
Induced edges, 24
Infectious diseases, 59
Inflection point, 79
Inherent edge, 24
Inhibited growth model, 78–81, 133
limitations, 81
Inputs, in habitat models, 179
Insecticides, 61
Insects, 5
control by cavity-nesting birds, 148
Instantaneous time, 36
Integrated forestry and wildlife management,
101. *See also* Integrated management
aquatic coordination and restoration
measures, 158–159
conflicts and competition, 143–145
engineering coordination and restoration
measures, 156–157
forest types, 101–114
grazing systems, 143
integrating factors, 136–146
land use coordination and restoration
measures, 158
landscape factor, 140–141

log-normal distribution, 138–139
National Research Council and, 145–146
natural rotation, 137–138
nonadjacency constraint, 139
range coordination and restoration measures,
155–156
recreation coordination and restoration
measures, 157–158
restoring/maintaining resources and diversity,
141–159
scale considerations, 137
stand size and distribution considerations, 138
timber coordination and restoration measures,
153–155
time considerations, 137
tradition and, 141–143
watershed coordination and restoration
measures, 157
wildlife coordination and restoration
measures, 151–153
Integrated management, 194–195. *See also*
Integrated forestry and wildlife
management
adaptive management, 195–196
systems approach, 195
Integrated resource management (IRM), 204–205
Integrated resource management planning
(IRMP), 163, 194, 204
Intergovernmental Panel on Climate Change
(IPCC), 46
Interspecies disease transmission, 59, 60
Interspersion, xiv, 20
Introduced species, 4
Invader species, 191, 192
Inventory, 205
categories, 206–207
geographic information systems, 209
intensity of, 207–208
map scale, 208–209
multiresource, 205–206
use of nested plots, 210
Invertebrates, eaten by animals, 238
Irruptive fluctuations, 84, 85
Island archipelago approach, 176

J

Juxtaposition, xiv, 20

K

K-selection, 67

L

Lakes, 13–14
Land rehabilitation, 23

Land use, coordination and restoration measures, 158
Landscape diversity, 22
Landscape ecology, 22
Landscape factors, 140–141
Landscapes, 164
Larch forests, 112
   common wildlife species, 112
   in Western states, 104
Large mammals, determining relative abundance, 217
Law of dispersion, 23
Law of the minimum, 37
Law of toleration, 36, 37
Laws and policy, 197
   Endangered Species Act, 198
   Fish and Wildlife Coordination Act, 197
   Forest and Rangeland Renewable Resources Planning Act, 199
   governing policies, 199–201
   Multiple-Use, Sustained-Yield Act, 198
   National Environmental Policy Act, 198
   National Forest Management Act (NFMA), 199
   Sikes Act, 198
Lead poisoning, 60
Legislation, protecting predators, 62
Length conversion factors, 255
Lentic water, 12
Leopold report, 61–62
Life, 3
Life equations, 132
   for wild turkey, 132
Life form, 5, 170, 171–172, 238, 246
   animals, 5
   plants, 10–11
Life history
   animals, 5
   plants, 11
Life-Needs table, 225, 246
   in FAAWN, 245–246
   forest components, 21
Life table, 82–83
   for tassel-eared squirrel, 83
Light tolerance, 115–116
Limiting factors, 37
Limnetic zones, 14
Lincoln index, 215
Links, in relational systems, 231–232
Littoral zone, 14
Livestock
   animal unit equivalents, 144
   conflicts with wildlife, 142
   food habits overlap with wildlife, 143, 144
Livestock grazing, 90, 114
   overuse of streamside vegetation by, 136
   on public lands, 142

Loblolly-shortleaf pine forests, 103, 108
   common wildlife species, 108
Local areas, 238
Local ecosystems, 164
Local populations, 70, 71
Lodgepole pine, 112
   common wildlife species, 112
   in Western states, 104
Log-normal distribution, 138–139
Logical domain, 230
Long rotations, 176
Long-term adaptability, 72
Long-term plans, 202
Long-term productivity, 44
Longevity, 66, 67
   physiological, 67
Longleaf-slash pine forests, 103, 106, 108
   common wildlife species, 108
Loss of habitat, 137
   for big game species, 65
   effects of roads, 58
   through fire, 56
Loss of vegetation, through fire, 56
Lotic water, 12
Lotka-Volterra equations, 62–63
Low-intensity inventories, 207
Low-mobility species, 67
   effects of roads on, 58
Low-value food, 87
Luangwa Valley project, xiv
Lyme arthritis, 60
Lyme disease, 60

M

Malentities, 6
Malnutrition, due to weather, 59
Mammals, 5
   averages for home range, 91, 92
   determining relative abundance, 217
Man-made hazards, 125
Management by objectives (MBO), 202
Management goals, dollar values required, 183
Management indicator species (MIS), 173
Management strategies. See Ecosystem management strategies
Management unit inventories, 206
Management units, 145
   required sizes, 163, 165
Manipulative functions, in habitat models, 179
Many-to-many relationships, 231
Map scales, 208–209
Maple-beech-birch forests, 103
Mates, 6
Matrix, 23
Maximum sustained yield (MSY), 133
   limitations, 133–134

Mean annual increment (MAI), 41, 122
Mean annual temperature, rise in, 46
Meandering edges, 24
Menominee paradigm, 47–48
Menominee tribe, 47
Metapopulation, 23, 70, 71
Microhabitat, 9
Microsites, 8, 19
    biological, 20
    in FAAWN data model, 238, 245–246
    nests as, 89
    physical, 20
Microsoft Access, 225, 245, 249, 251
    Filter menu, 245
Migration routes, blockage by fences, 58
Minimum viable populations, 71
Mitigation, 142
Mixed-age forests, 4
Mobility
    and loss of habitat, 137
    low- vs. high-, 67
    reproductive characteristics associated with,
        93
    as resource for survival and reproduction,
        91–92
Moderate-value food, 87
Monoclimax, 38
Monogamous species, 66
Mortality, 70
    dependence on population density, 80
Multiple-Use, Sustained-Yield Act, 198
Multiresource inventory, 205–206
Multiresource objectives, 115
Multispecies cycles, 84
Multispecies wildlife management, 48, 141,
        169–170
Muskrats, mink predation of, 62

N

Natality, 70
    and mortality, 80, 82
National Academy of Science (NAS), 146
National Atlas map, 27
    forest types, 102
National Cartographic Information Center, 208
National Environmental Policy Act, 136, 198
National Forest Management Act (NFMA), 16,
        71, 119, 199
National Forests
    acres in, 234
    by state, 107–108, 236
National inventories, 206
National Research Council, 145–146
National table, in FAAWN, 248
Natural events, 44–45
Natural populations, 70

Natural rotation, 137–138
Natural terrain, 140
Needle leaf trees, 10
Nested plots, 209
    in habitat inventory, 210
Nested questions, 230
Nesting cover, 88, 246
Net reproductive rate, 74
Neural network programs, 227
Nonadjacency constraint, 139
Nongame species, 4, 174
Noninfectious diseases, 60
Nonprotected species, 4
Nonrenewable resources, 42
Northern Arizona University, xix
Nutrient cycles, 31, 32–33
Nutrient pathways, 34–35

O

Oak-gum-cypress forests, 103, 109
    common wildlife species, 109
Oak-hickory forests, 103, 109
    common wildlife species, 109
Oak-pine forests, 103, 108
    common wildlife species, 108
Ocean-basin ecosystems, 164
Ocean pH, 46
Oceans, warming of, 46
Old forest conditions, 41–42, 120–121
    preserving, 138
    stand diversity in, 120
Old growth, 40
Old-growth acre, 41
Old-growth forests, 4
Oligotrophic states, 13
One-to-many relationships, 231
One-to-one relationships, 231
Open nests, 149
Open water zones, 14
Openings, 18
    in stands, 28
Operational plans, 202
Order of magnitude, 93–95, 237
    in acres, 94
    comparing home range to size of area, 94
Outputs, in habitat models, 179
Overgrazing, 144
Overstory, 25
Oxygen, in cold water, 13
Ozone theory, 84

P

Patch, 23
Pathogens, 59
Pattern recognition, 183–184, 227

Patton, David, xv, xix
Pecking order, 69
Pellet count technique, for estimating deer and
    elk populations, 217
Persistence, 35
Pesticides, 61, 126
Peterson index, 215
pH range
    changes in oceans, 46
    of water, 12
Physical components, 20, 31
    identifying through relational theory, 225
Physical domain, 230
Physical features sampling, 213
Physical microsites, 20
Physiognomy, 24
Physiological longevity, 67
    uninhibited growth model, 73
Pinyon-juniper forests, 113
    common wildlife species, 113
    in Western states, 104
Pioneering populations, 73
Planning, 196–197
    background information checklist, 203
    decision support systems, 203–204
    factors to consider, 201–202
    laws and policy, 197
    operational plans, 202
    project implementation, 204–205
    rationale, 201
Plant code, 239
Plant communities, change over ecological time,
    2
Plant growth response, 82
Plant nutritional value, index of, 87
Plant species, dependence on sun, 2
Plant species composition, 190
Plant succession, 37
Plant value, 191
Plantation effect, 121
Plants
    autotrophic food production, 33
    dependence of animal species on, 134
    in forest environment, 10–11
    life form, 10–11
    life history, 11
Plants table, in FAAWN, 248
Plants-Types table, in FAAWN, 248
Plot-based sampling, 210
Plot size, 211
Plotless sampling, 210
Poaching, 65–66
Point-based sampling, 210
Poisons, 128
Policy statements, 199–200
    changing interpretations of, 200–201
Pollution, 90

Polyclimax theory, 38
Polygamy, 66
Ponderosa pine, 31, 111–112, 148
    common wildlife species, 111–112
    coyote seasonal food habits in, 86
    ECOSIM model, 170
    nest zone, tassel-eared squirrel, 89
    in Western states, 104
Population change
    litter size change and, 78
    percentage change and exponential rate, 76
    rates of, 74–78
    stable age distribution, 77
Population characteristics, 69–72
    density, 71
    distribution, 71
    levels of protection, 72
Population density, 71
    of prey and predators, 64
Population distribution, 71
Population dynamics, 69
Population equilibrium, 78
Population manipulation, 124
Population regulation, 80
Populations, 30
Precipitation, young quail number relative to, 90
Predator control, in wildlife management,
    127–128
Predator-prey relationships, 62, 63–64, 84
Predators, 6, 61–64
    general propositions, 63–64
    Leopold report, 61–62
    Lotka-Volterra equations, 62–63
Preferred food, 87
Prescribed fire, 135
Present time, 36
Primary cavity nesters (PCNs), 148
Primary habitat, 8
Primary succession, 38–39
Principle of population, 85
Probabilities, in pattern recognition, 183
Processes, in ecological systems, 31
Producer-consumer biomass, 31
Productivity, long-term, 44
Professional hunters/trappers, 128
Profundal zone, 14
Project implementation, 204–205
Promiscuity, 66
Protection, hierarchical levels of, 72
Protection levels, 71, 72
Protein content, of riparian vegetation, 15
Protein requirements, of wild animals, 87
Public lands
    grazing rates on, 145
    livestock use, 142
    removal of cattle from, 145

Q

Quality control, in database entry, 228
Quantified habitat models, 190
Quantitative models, 184

R

R-selection, 67
Rabies *(Rhabdovirus)*, 59
Rainfall, 90
Rainstorms, 58
Random samples, 212
Random theory, 84
Range conditions, 203
Range coordination and restoration measures, 155–156
Range ecosystems, 25
Range management practices, xiii
Rangeland species, 4, 142
Rates of extinction, 22
Reaction to disturbance, 37
Recreation coordination and restoration measures, 157–158
Recreation facilities, 203
Recreational campground use, 136
Reductionists, 39
Redwood forests, 113
    common wildlife species, 113
    in Western states, 104
Refuges, 130
Regional code, 239
Regional inventories, 206
Regional names, 239
Regions, and forest types, 101
Regions linking table, in FAAWN, 247
Reintroduction of species, 64
Relational data model, 21
Relational database systems, 228
    design criteria, 230–231
    design evaluation and testing, 232
    design trade-offs, 232
    links determination, 232
    many-to-many relationships, 231
    one-to-many relationships, 231
    one-to-one relationships, 231
    quality control, 228
    relation theory, 228
    relationship systems, 229–230
    software rules, 231
    structure and organization, 228
    validation and testing, 233
Relational databases, 225, 228
Relational systems, 231
Relational theory, 225, 228

Relations, 229
    in FAAWN data model, 233
    FAAWN directory, 234–243
Relationship systems, 229–230
Renewable resources, 42
Reproduction, 11. *See also* Survival and reproduction
    biology of, 66–68
    low-mobility *vs.* high-mobility species, 67
    and physiological longevity, 67
    requirements, 5
Reptiles, 5
    effects of fire on, 56
Resilience, of ecosystems, 39
Resolution, levels in habitat models, 178
Resource base, 18
Resource maintenance, 146–159
Resources, 6, 85–86
    food, 86–87
    mobility, 91–92
    order of magnitude (Om), 93
    space, 91
    water, 89–91
    wildlife management considerations, 134–136
Rest-rotation grazing, 143
Resting cover, 88
Restoration, 152
    engineering measures, 156–157
    of land use, 158
    of range, 15–156
    of timber, 153–155
    of wildlife, 151–153
Riparian meadow habitats, and elk, 58
Riparian vegetation, 14, 15
    at high elevation, 16
    loss of Southwestern, 65
Riparian vegetation corridors, 176
Riparian zones, 4, 15–16
Risk analysis, 71, 72
Roads
    as man-made hazards, 125
    as wildlife hazards, 57–58
Roosting cover, 88, 246
    snags as, 149
Rotation, 122–123
Running water, 12
RUNWILD data model, 225

S

Salmon, 135
    human actions affecting, 136
Salvage cuts, 121
Sample units, shape and size, 211–212
Sampling, 209
    circular plot sizes, 211
    collectible information, 213–214

counting signs, 216–218
determining relative mammal abundance, 217
direct animal counting, 215–216
estimating animal numbers, 215
nested plots, 210
pellet count technique, 217
sample unit shape/size, 211–212
sampling design, 212–213
tree measurements, 218–219
types of, 210
uses of habitat data, 214–215
Sampling design, 212–213
Sampling systems, 210
Satellite imagery, 142, 203
Scale, 165
    integrated management considerations, 137
Scientific names, of animals and plants, 240
Sea level rise, 46
Seasonal food habits, coyotes in AZ ponderosa
    pine forests, 86
Second law of thermodynamics, 34
Secondary habitat, 9
Secondary succession, 39
Seed tree harvesting, 117, 120, 122
Seeps and springs, 14–15
Selection harvesting, 117, 120
Sensitivity analysis, 179
Seral stages, 39
Sere, 37
Severe weather conditions, impacts on wildlife,
    58–59
Sexual maturity, age of, 66
Shade, tolerance of, 115–116
Shantz-Zon map, 27
Shelterwood harvesting, 117, 119, 122
Short-term adaptability, 72
Shrubland, 25
Shrubs, 10
    food and cover uses, 150
    planting and production of, 150
Sign counting, 216–218
Sikes Act, 198
Silvicultural systems, 115
Single species management, 166, 168–169
Small mammals, determining relative
    abundance, 217
Snag density, 149
Snags, 1, 28–29, 138, 148
    food and cover uses, 149
    management for cavity-nesting birds, 170
    use by wildlife, 149
    wildlife uses, 29
Snow, 58
Snow movement, control methods, 124
Sociability, 25
Society of American Foresters (SAF), task force, 41
Soil, 35

Soil erosion, 191
Soil fertility, and animal density, 35
Soil loss, 65
Soil properties, 190
Soil surveys, 203
Solar ecosystem, 164
Space
    in ecological systems, 36
    as resource for survival and reproduction, 91
    territory as expression of, 68
Spacing, homogeneous/random, 25
Species diversity, 22
    maintaining through predation, 63
Species grouping, based on common habitat
    requirements, 169
Species richness, 17
Specific heat, of water, 12
Spiny-rayed fish, 135–136
Spotted owl, old forest habitat, 120–121
Spring turnover, 13
Spruce-fir forests, 103, 106
    common wildlife species, 106
Squirrels, escape cover, 88
Stability
    in climax patterns, 39
    and complexity, 16, 35
    of ecosystems, 3
    of wildlife populations, 124
Stable age distribution, 76, 77
Stable populations, 83
Stand density index (SDI), 185, 186, 219
Stand distribution, 138
    and wildlife management, 139
Stand initiation, 40
Stand size, 123, 138
    following log-normal curve, 138
    and landscape factor, 140
    and wildlife management, 139
Stand structural stages, 18
Stand structure model, 186
    matrix, 187
Stands of trees, 23, 28, 93
    cover value, 121
    defining for ecosystems management, 164
    documenting by aerial/satellite imagery, 142
    openings in, 28
Staple food, 87
State within region, 101
States table, 240–241
    in FAAWN, 246–247
Stem exclusion, 40
Stocking
    of game animals, 128
    rationales for, 129
Storm intensity, changes in, 46
Strategic plans, 202
Stratified samples, 212

Streams, 14
  current and velocity, 14
Stress theory, 84
Structural layers, 18
Structure, in ecological systems, 31
Structured query language (SQL), 225
Stuffing food, 87
Stumps, 29
Succession, 43, 194
  direct techniques for modifying, 135
Successional position, 191
Sunspot theory, 84
Suppressed trees, 25
Survival and reproduction, 55
  biology of, 66–85
  cognitive map of factors, 2
  diseases and, 59–61
  environmental factors affecting animals', 6
  in FAAWN data model, 241
  fire hazards, 55–57
  food/cover designations for, 245
  hazards to, 55–59
  human roles, 64–66
  predators and, 61–64
  resources and, 85–95
  role of habitat, 7
Sustainability, 1, 44, 45, 163
Sustained yield, 47
  in game animals, 133
Synecology, 39
Synthetic descriptors, 20
Systematic samples, 212
Systems approach, 166, 195

T

Tables, 229
Tassel-eared squirrel, 137
  composite life table, 83
  forest type habitat, 102
  habitat capability model, 169
  nest zone in AZ ponderosa pine, 89
  nesting cover for, 89
Temperature
  conversion factors, 256
  fish sensitivity to, 13
Tension zones, 24
Territory, 68–69
Thermal cover, 88, 246
Thermal stratification, in water bodies, 13
Thermocline, 13
Thermodynamics, laws of, 34
Thinning, 121
Threatened species, 4
  controversies, 115
Ticks, 59, 60, 126
Timber coordination and restoration, 153–155

Timber harvesting, xii, 135. *See also* Tree harvesting
  food and cover remaining after, 141
  mitigation measures, 142
  number of trees remaining, 136
Timber inventory records, 203
Time
  in ecological systems, 36
  integrated forest-wildlife management considerations, 137
Time perspective, for ecosystems management, 163
Tornadoes, 58
Toxic chemicals, 59
Tradition, considerations in integrated management, 141–143
Transects, 210
Transparency, of water, 12
Trapped wild animals, 130
Tree density, 185
Tree diameter, 219
Tree harvesting, 114, 136. *See also* Timber harvesting
  checkerboard patterns, 139
  conflict with provision of habitat, 141
  maximum size limitations on, 139
  stand regeneration after, 115
Tree measurements, 218–219
Tree spacing, 121
Tree species tolerance, 115–116
Tree stand canopy, 239
Trees
  dead, 28–29
  defined, 10
  food and cover uses, 147
  life history, 11
  stands, 28
  structural stages, 39–40
Trend surveys, 203
Tropical forests, 101
Trout
  expense of habitat improvement, 159
  human actions affecting, 136
Tumors, 60
Turbidity, of water, 12
Turkeys
  roosting cover for, 88
  sample data with habitat attributes, 184
Turnover, in water systems, 13
Types-Codes table, in FAAWN, 248
Types-States table, in FAAWN, 248
Types within state, 101

U

Umwelt perspective, xiii, 7
  animal *vs.* human views of vegetation, 8

U.N. Food and Agriculture Organization, xiv
Understory, 25
    reinitiation, 40
UNESCO vegetation classification system, 25
Uneven-aged stands, 116, 117
    cutting cycle in, 123
    forest regulation to produce, 123
Uninhibited growth model, 73–74, 133
Unoccupied habitats, populating through
        stocking, 129
Upland vegetation, 15
Urbanization, 65
    loss of habitat to, 65
U.S. Department of Agriculture, 146
U.S. Department of Interior, 146

V

Value, human viewpoint, 3
Vegetation, 24–25
    animal *vs.* human views, 8
    classification, 25–27
    coarse woody debris, 29–30
    condition of, 191
    dead trees, 28–29
    loss through fire, 56
    openings in stands, 28
    snags, 28–29
    stands of trees, 28
Vegetation sampling, 211, 213
Vegetation structures, and cover for white-tailed
        deer, 88
Vegetation type maps, 203
Vegetation types, wildlife associated with, 230
Vegetation zone, 26
Vehicle collisions, with animals, 57
Vertebrates, eaten by animals, 242
Vertical diversity, 17
    in forest ecosystem, 18
Viable populations, 22, 71
Viral diseases, 59
Viscosity, of water, 12
Volume, 18
    conversion factors, 255
Volume regulation, 121

W

Warm water, 13
Warts, 60
Waste matter, 34
Water
    density, 12
    in forest environment, 11–12
    and freshwater habitats, 12–13
    lakes, 13–14
    pH range, 12

properties, 12
    quail number relative to precipitation, 90
    requirements by animal species, 90
    riparian zones, 15–16
    role in survival and reproduction, 89–91
    seeps and springs, 14–15
    specific heat, 12
    streams, 14
    transparency, 12
    turbidity, 12
    viscosity, 12
Water current, 13
Water inventory, 203
Water map, 215
Water requirements, by animal species, 90
Water skin, 12
Watersheds
    coordination and restoration measures, 157
    management and protection of, 135
Weather, 45
    as wildlife hazard, 58–59
Webs, xiii, 6–7, 91. *See also* Food webs
    and indirect human effects on wildlife, 64
    similarities to data models, 227
Weight conversion factors, 255
Western hardwoods forests, 114
    common wildlife species, 114
    in Western states, 104
Western states, forest cover types, 102, 104,
        110–114
Western white pine, 112
    common wildlife species, 112
    in Western states, 104
Wetland drainage, 90
Wetland species, 4
White-red-jack pine forests, 1–3, 103–104, 106
    common wildlife species, 104, 106
White-tailed deer
    diseases, 126
    as high-mobility species, 67
    illegal hunting, 65
    malnutrition due to weather, 59
    protein requirements, 87
    vegetation structures and cover type for, 88
Wild, 3
Wild turkey
    life equation, 132
    protein requirements, 87
Wilderness species, 4
WILDHARE data model, 225
Wildlife
    animal life history, 5
    animal unit equivalents, 144
    classification, 4
    defined, 3–4
    food and cover uses of dead trees and snags, 149
    food habits overlap with livestock, 144

in forests, 3
importance of forest ecosystems to, 115
indirect effects of fire on, 56
life form, 5
Wildlife census data, 203
Wildlife cover, 88
Wildlife data models, 186–188
Wildlife environment, 5
centrum and web, 6–7
habitat, 7–10
Wildlife habitat relationships, 166–168
featured species management, 170–171
guilds, 171
life form, 171–172
multispecies management, 169–170
single species management, 168–169
Wildlife management, 124, 146. *See also*
Integrated forestry and wildlife
management
biology in, 128–134
diseases, 125–126
hazards, 124–125
human factors, 127
integrating with forestry management, 101
predator control, 127–128
resource considerations, 134–136
Wildlife Management Techniques Manual, 83
Wildlife observations sampling, 214
Wildlife occupancy sampling, 214
Wildlife restoration measures, 151–153
Wildlife species
in aspen-birch forests, 110
characteristics in database, 229
in Douglas-fir forests, 111
in elm-oak-cottonwood forests, 110
in fir-spruce forests, 113
in hemlock-Sitka spruce forests, 111

in larch forests, 112
in loblolly-shortleaf pine forests, 108
in lodgepole pine forests, 112
in longleaf-slash pine forests, 108
in oak-gum-cypress forests, 109
in oak-hickory forests, 109
in oak-pine forests, 108
in pinyon-juniper forests, 113
in ponderosa pine forests, 111–112
in redwood forests, 113
in spruce-fir forests, 106
in western hardwoods forests, 114
in western white pine forests, 112
in white-red-jack pine forests, 104, 106
Windbreaks, 124
Woodland, 25, 26
Woody debris, 29–30
Working concepts, 1
Menominee paradigm, 47–48
common knowledge axions, 1–3
complementarity principle, 48
forest diversity, 16–24
forest plants, 10–11
forest succession, 37–44
forest wildlife, 3–5
forests as ecological systems, 30–37
healthy forest conditions, 44–47
vegetation, 24–30
water in forest environment, 11–16
wildlife environment, 5–10

Z

Zip-code, 247
in FAAWN data model, 243